Praise for *Beyond Darwin*

"At a moment when the world feels like it's falling apart, Bruce delivers the scientific proof that this chaos isn't the end—it's the beginning of something extraordinary. This book doesn't just challenge old thinking. It gives us a road map for conscious evolution built on cooperation, not competition. If you want to understand the science behind why transformation is not only possible but inevitable, read this book."

—Tony Robbins, #1 *New York Times* bestselling author,
speaker, and philanthropist

"In *Beyond Darwin*, Dr. Bruce Lipton beautifully bridges science and consciousness in a way that is both illuminating and deeply hopeful. He reveals that cooperation—not competition—is the deeper intelligence guiding evolution, and that when we awaken to our interconnectedness, we become conscious participants in the healing and evolution of our planet."

—Anita Moorjani, *New York Times* bestselling author of *Dying to Be Me*

"For thirty years, I have taught what Bruce Lipton now proves: the body is not a machine to be managed but a living field to be awakened. *Beyond Darwin* gives science to what the ancients always knew. Cooperation is not idealism. It is biology. This book is medicine for our moment."

—Qigong Master Mingtong Gu, author of
Coming Home to Embodied Awakening

"A life-saving and heartwarming rehabilitation of traditional and often maligned concepts that rescue the emergent sciences of life from an early demise in the jaws of obsolescence."

—Mária Sági, PhD, psychologist and author of *Remote Healing*

"Bruce Lipton's new book opens a long-awaiting era in which new sciences become relevant to human well-being and even survival on a small planet. Threatened by outdated thinking and misinformation, henceforth science becomes not only relevant but essential to mastering the basic knowledge without which there is no coherent understanding of who we are and how we relate to human life. A mind- and paradigm-shaking and reforming contribution to the evolution of humankind to be treasured for ages and generations to come."

—Ervin Laszlo, author and founder of The Club of Budapest

"The healing work I have witnessed across fifteen thousand certified practitioners in over one hundred countries tells me the same story this book tells: we are not accidents of biology locked in a struggle for survival. We are participants in something magnificent—a living garden meant to be tended with care, intention, and love.

If you have ever sensed that there is more to your health, your relationships, and your purpose than the old story has allowed, *Beyond Darwin* will give you the scientific language, the historical context, and—most importantly—the hope to step fully into that larger truth."

—Dr. Bradley Nelson, national bestselling author of *The Emotion Code*, *The Body Code*, and *The Heart Code*

"Understanding the root cause of a problem is the first step in solving it . . . Switching future progress from competition to cooperation provides real hope for our species. This is an important book. We are running out of time."

—David Hanscom, MD, orthopedic spine surgeon and author of *Calm Your Body, Heal Your Mind*

"Bruce helps us realize that evolution is something we can consciously contribute to and guide toward greater positivity and a healthier humanity. Such an empowering message we all need to hear and, above all, live by."

—Deva Premal and Miten, Grammy Award–nominated musicians

"In *Beyond Darwin*, the brilliant biologist and visionary Dr. Bruce Lipton takes a well-aimed hatchet to orthodox theories of genetics, survival-of-the-fittest evolution, and a worldview of life as predatory, self-serving, and solitary. This bold paradigm buster is filled with scientific revelations, signposting a life-enhancing vision of what it means to be human and a hopeful future based on conscious evolution and deep interconnection—just at the moment in our history when we most need it."

—Lynne McTaggart, international bestselling author of *The Field*

"*Beyond Darwin* is a bold and deeply integrative work. By weaving epigenetics, quantum physics, and fractal geometry into evolutionary theory, Dr. Lipton challenges the reigning narrative of survival of the fittest through a new, science-based theory of cooperation that shines a powerful light on the power of our humanity."

—Lauren Monroe, healing artist, creator of Mind Body Drum,
and cofounder of the Raven Drum Foundation

"At a time when many feel society is unraveling, *Beyond Darwin* offers a solution backed by science. Dr. Lipton's exploration of consciousness as a shaping force in reality is both provocative, meaningful, and deeply rooted in research. This book is arriving just when mankind needs it the most."

—Rick Allen, drummer for Def Leppard
and cofounder of the Raven Drum Foundation

"Dr. Bruce Lipton marshals decades of pioneering research to challenge the Darwinian dogma of random mutation and blind chance, replacing it with an empowering story of conscious evolution and human purpose. If you're ready to trade 'accidental existence' for something far more extraordinary, this is the book that will change how you see yourself and the world."

—Dean Radin, PhD, chief scientist at the Institute of Noetic Sciences
and author of *The Science of Magic*

"Integrating insights from systems science, cell biology, and quantum physics, Bruce Lipton masterfully weaves empirical data to open our minds beyond materialistic determinism, toward a more empowered future. Through sharing his own healing journey, Lipton brings authenticity and humanity to the science—reminding us that conscious participation lies at the heart of both personal and planetary healing."

—Dr. Shamini Jain, author of *Healing Ourselves*

"At a time when many feel society is unraveling, *Beyond Darwin* offers a solution backed by science. Dr. Lipton's exploration of consciousness as a shaping force in reality is both provocative, meaningful, and deeply rooted in research. If you've ever wondered what truly shapes our world, this book is your invitation to explore."

—Michael Bernard Beckwith, founder and CEO
of Agape International Spiritual Center

Also by Bruce H. Lipton, PhD

The Honeymoon Effect
Spontaneous Evolution
The Biology of Belief

BEYOND DARWIN

HOW EPIGENETICS, QUANTUM SCIENCE, AND COOPERATION SHAPE HUMANITY'S FUTURE

BRUCE H. LIPTON, PhD
WITH PATRICIA KING

BenBella Books, Inc.
Dallas, TX

BenBella

BenBella Books, Inc.
8080 N. Central Expressway
Suite 1700
Dallas, TX 75206
benbellabooks.com
Send feedback to feedback@benbellabooks.com

BenBella is a federally registered trademark.

Printed in the United States of America
10 9 8 7 6 5 4 3 2 1

Library of Congress Control Number: 2026013751
ISBN 9781637749364 (hardcover)
ISBN 9781637749371 (electronic)

Copyediting by Scott Calamar
Illustrations by Ralph Voltz
Proofreading by Cheryl Beacham and Jill Kramer
Indexing by Elise Hess
Text design and composition by Jordan Koluch
Cover design by Paul McCarthy
Cover image © Adobe Stock / Nataliia Yudina
Printed by Lake Book Manufacturing

This book is dedicated to Mother Nature.

Evolutionary science emphasizes that Nature is truly the Mother of us all, for we have evolved from her Garden. Indigenous populations have always seen their role as Gardeners who tend to the environment, but later civilizations have instead been dedicated to dominion over Mother Nature, which has led to an impending sixth Mass Extinction Event. The first Earth Day, April 22, 1970, offered the powerful message that humanity needs to step back from that looming Mass Extinction event by protecting and restoring the Garden.

While forces in our current civilization, including corporate and political interests, have thwarted the efforts inspired by every Earth Day since 1970, it is my dearest hope that the information offered in this work will re-instill in Humanity a fervor to stop environmental desecration by embracing our mission to maintain our one precious Garden.

Contents

PART 2

Foreword

Sometimes we find the truth of what "is" in our lives by first discovering what "is not." Through the process of elimination, we eventually zero in on the understanding we're searching for. From our personal relationships with family, friends, and lovers, to the war and peace between nations, we seem to learn the great lessons of life in precisely this way. In addition to obvious relationships in our everyday lives, this principle also applies to the deepest questions of life itself, such as who we are and where we come from.

Since the 1859 publication of Charles Darwin's book *On the Origin of Species*, where he attempted to describe the nature of evolution, scientists and researchers have searched for evidence to better understand what's true, and what's not true, when it comes to Darwin's theories. The explosion of discoveries resulting from the search has left little doubt that many long-standing views about life, our bodies, and Darwin's theory itself, have to change. The reason is simple: *The ideas are wrong.*

For example, if we could summarize some of the new discoveries that no longer support the old ideas, what would they tell us? What does the best science of our time indicate about who we are, and who we are not? The partial list that follows gives us an idea as to where the New Science is headed. It's a fact that:

- The theory of living cells mutating randomly (evolving) over long periods of time does not explain the origin or complexities of human life.
- The biological link between humans and earlier humanlike life-forms in our ancestral tree is inferred, speculative, and not proven.
- DNA studies prove that we anatomical modern humans (AMH) did not descend from Neanderthal families, as previously believed.
- The DNA that makes us human and gives us our uniqueness could not have formed in the ways that it has, purely from natural, long, slow gradual processes of random mutations and lucky biology.

So now that we know some of the things we're "not," what does the best science of our time tell us about who we *are?* The answer to this question is revealed within the sixteen chapters that you hold in your hand.

THE NEW HUMAN STORY

Through decades of pioneering research by stem-cell biologist Dr. Bruce Lipton, along with more recent studies by other scientists, the science of epigenetics—the influence of the environment on our genes—is now percolating into mainstream textbooks that we rely on for health and well-being.

These discoveries have overturned 150 years of Darwinian thinking when it comes to us—our origin, our potential, and perhaps most importantly, our ultimate destiny. The undeniable evidence of epigenetic influence is forcing the scientific community to rethink the most basic beliefs that lie at the core of the decisions we make and the way we live. This is where this book, *Beyond Darwin: How Epigenetics, Quantum Science, and Cooperation Shape Humanity's Future* comes in.

The pages that follow are the keys that unlock a way of thinking that radically changes the way we've been led to view ourselves in the past. I invite you to embrace the discoveries and explore what they mean to you. Talk them over with the important people in your life; and discover if, and how, they change the story of your existence, as well as the way you respond to the challenges of everyday life.

FOREWORD

THIS BOOK

When the evidence no longer supports the story, it's time for a new story. *Beyond Darwin* is written with one purpose in mind: to empower you with a new scientific story of who you are and the extraordinary potential that awaits as you accept the deep truth of your potential to thrive in life. From scientists and engineers to homemakers and healers, Bruce Lipton's rock-solid discoveries provide the perfect blueprint for an inspired life, healthy relationships, and a healed world.

We're better people, and our lives are more fulfilled, as a result of Lipton's courage to share what he's discovered, and his passion to lead us to a world where cooperation replaces competition, and a sharing economy replaces the thinking of scarcity and limitations.

The question we face is simple: Do we love ourselves enough to move beyond Darwin's ideas, to accept the new human story that science is telling us? This book, *Beyond Darwin*, gives us everything we need to answer this question for ourselves, and to live the blessing of our answer in our everyday lives.

—Gregg Braden, scientist and
New York Times bestselling author of books including
Pure Human and *The Divine Matrix*

Prologue

THE END IS NIGH

Whether you are following the news, surfing the web, or even looking out your own window, it is apparent that the planet is caught up in a wave of global chaos. Ranging from political, to social, to economic, to racial and religious upheaval, civilization is in a state of disharmony.

Seemingly disconnected, all the chaos is driven by an underpublicized reality that is undermining all of civilization: Our current civilization is unsustainable. The unappreciated fact is that society is living beyond Earth's available resources. To maintain itself, humans have pillaged the planet's resources, destabilized the environment, polluted the air, the soil, and the ocean. In the last fifty years, human behavior has led to the loss of over two-thirds of the planet's animal population.

A major contribution to the problem is that human civilization holds the belief that we are a separate, independent entity from "Nature." This belief was first introduced to Western civilization in the Bible's chapter Genesis, wherein God first creates Nature and then later introduces humans as a separate act of creation. The story goes on to acknowledge that humans have dominion over the planet. Later, Darwinian science's view of evolution suggested that humans arose because of a series of accidental genetic mutations. If we are the result of random alterations in heredity, the implication is that our appearance on Earth was simply a matter of chance, with no apparent reason for our existence.

Our belief in a purposeless dominion over the planet is destroying the web of life to the extent that we are now facing a mass extinction event. Science recognizes that we are on a course that will result in an "irreversible" collapse of civilization within the next twenty years. We are running out of time to change the course of our fate. While the notion of a collapse of civilization seems to many to be a radical assumption, the simple fact is that Western civilization has already experienced the rise and fall of three previous versions of civilization.

Ten thousand years ago, the civilization of indigenous humans was living in an era called *animism*, in which every material structure from a drop of rain to a blade of grass, to a tree, and up to a human was considered to be both a physical structure and a spiritual entity.

Subsequently, around 4,000 BC, animism evolved into a new version of civilization called *polytheism*, which flourished in Greece, Rome, and Egypt. In these civilizations, physical objects lost their spiritual character because "Spirit" was attributed to a pantheon of gods that lived on Earth (Mount Olympus).

Around 800 CE (formerly referred to as "AD"), the polytheistic societies fell and were replaced by a new version of civilization, *monotheism*, the belief in a single God, a spiritual source that no longer lived on the planet. It was in this era that the Church of Rome, claiming to be God's representative on Earth, controlled the fate of Western civilization.

After 1859, when Darwin published his theory *The Origin of Species by Means of Natural Selection*, civilization adopted the scientific view that life was "controlled" by physical genes. The Church's control based on the philosophy of monotheism gave way to a new and current epoch, *scientific materialism*, which focuses on the material realm and excludes spiritualism.

Once, the technological advances in the West were the primary influences impacting planetary health. However, over the last century, with globally shared technological evolution and advances in travel and communication, the formerly disparate Western and Eastern versions of civilizations have since merged into a unifying planetary civilization.

Currently, we must consider the collective actions of all human societies across the globe in how their activities are influencing the fate of the planet.

In my 2009 book *Spontaneous Evolution*, I described how the character of each civilization is determined by the answers to a set of perennial questions about the origin of humans and our relationship to the planet. There are three fundamental perennial questions:

1. How did we get here?
2. Why are we here?
3. Now that we are here, how do we make the best of it?

The character of each of the four versions of Western civilization is shaped by the way they answered the questions:

Animism: (1) *How did we get here?* Mother Earth and Father Sky. (2) *Why are we here?* To tend the Garden. (3) *How to make the best of it?* Live in balance with Nature.

Polytheism: (1*) How did we get here?* From chaos. (2) *Why are we here?* No reason, it was just an accident. (3) *How to make the best of it?* Don't anger the gods.

Monotheism: (1) *How did we get here?* Divine intervention. (2) *Why are we here?* To live out plays of morality. Are we worthy of entering Heaven? (3) *How to make the best of it?* Follow the laws of the Bible.

Scientific Materialism: (1) *How did we get here?* Random genetic mutations. (2) *Why are we here?* No reason; it was just genetic accidents. (3*) How to make the best of it?* Survival of the fittest in the struggle for life.

When the acceptable answers to the perennial questions change, those answers inevitably lead to a new version of civilization, one in which the new answers shape the character of our evolving civilization.

Over the last twenty to thirty years, modern science has experienced a renaissance in its fundamental beliefs. Once, mutations were attributed to accidental alterations in heredity. Today, the New Science of epigenetics reveals the nature of beneficial or environmentally adapted mutations that delineate a direct connection between an organism and the environment in which it lives. This is why every organism appears to be perfectly adapted to its own environment: Polar bears live in the Arctic and are not found in Africa, roses live in temperate environments and are not found in the Arctic, fish live in the ocean and are not found on land. Organisms and environment are in lockstep integration. Are humans genetic accidents? No, as evidenced by the science of epigenetic mechanisms, we are the consequence of environmental adaption. We arose as participants supporting the harmony of the Garden in which we evolved.

In a world based on Newtonian physics, material objects are distinct and separate from the invisible energetic realm.

This is why in today's world we do not emphasize the influence of energy, which includes consciousness and spirituality, in shaping the course of our lives. However, in the realm of quantum physics, the most valid of all the sciences, matter is an illusion because everything in the Universe is an expression of energy, as Albert Einstein stressed: "The *field* (energy) is the sole governing agency of matter." The New Science emphasizes the fact that the invisible realm of consciousness and spirit play a primary role in determining the character of our lives.

New research changes the previous answers to the perennial questions. Modern science now recognizes that cooperation, rather than competition, is the fundamental behavior to ensure our survival in an evolving world. The new answers to the perennial questions are as follows:

1. *How did we get here?* Cooperative interaction between humans and the environment in which we evolved.
2. *Why are we here?* Humans are an element of the garden we evolved in. A garden by definition is an expression of harmony among the community of organisms comprising that garden, not the Darwinian battleground we have been perceiving it as.
3. *How do we make the best of it?* As the indigenous people first recognized, our role is to be gardeners who help keep balance and harmony among all the organisms comprising the web of life.

The new answers to the questions reveal why our current civilization is in decline and that we are on the threshold of manifesting a new version of human behavior that can not only sustain us but help us thrive into the future.

The information offered in this book, based on epigenetics, quantum science, and cooperation, emphasizes a scientific and behavioral foundation for a new theory, *conscious evolution*, that can transcend the current self-destructive consequences of Darwinian theory and guide us safely into a sustainable future.

PART 1

INTRODUCTION

After I finished *The Biology of Belief* manuscript two decades ago, an editor at one East Coast book publisher said her company wouldn't publish it because a member of the committee that reviewed my book proposal thought I was a creationist! Since then, I've gotten used to being labeled incorrectly—I've given up trying to get Wikipedia to stop calling me a "pseudo scientist." Instead, I have focused on constructing my comprehensive theory of evolution and presenting it all over the world.

Over the years, the scientific evidence for my theory kept getting stronger with mind-altering quantum physics discoveries, epigenetic studies that confirmed my early cell-cloning experiments that convinced me that genes do not control biology, and complex fractal geometry studies powered by increasingly advanced computers that highlight the recurring patterns of evolution.

Also, gratifying for me, my audiences kept getting bigger, including in the last two years ten thousand in Basel, Switzerland, and the readership for *The Biology of Belief* has skyrocketed. *The Biology of Belief* has sold more than a million copies, has been translated into fifty languages, and is being used as a textbook for medical students in Argentina at the Universidad Nacional de Rosario Medical School.

I believe the science-backed message I deliver resonates strongly because people are hungry for hope in a world that seems terminally chaotic. And I do offer hope. I believe that the chaos my sometimes-despairing audience members cite signifies a necessary evolutionary step that is a harbinger of a better future. The scientific framework of the theory I offer has convinced me that despite lots of contemporary evidence to the contrary, the story of evolution is a hopeful one, a story of expanding consciousness and reassuring, repeated fractal patterns. Because we humans have agency, we can emerge from the chaos to the highest level of *conscious evolution* characterized not by survival-of-the-fittest competition but by cooperation.

In this book I offer a succinct (challenging for me!) version of my theory and the framework for the mechanics of evolution as I understand them. As always, my work is grounded in my knowledge of the cells I first peered at under a microscope in Mrs. Novak's second-grade classroom. Those cells launched my lifelong love affair with science and, though I couldn't have articulated it at the time, my search for a comprehensive theory of evolution as well—I've always been intrigued by the puzzle of how, where, and when the primitive reflexive consciousness of bacteria led to the powerful, creative self-consciousness of human beings, each of whom is made up of a remarkable collection of fifty trillion cells. I know that creative self-consciousness gives highly evolved humans terrifying power to destroy ourselves and other species, but I also know that it gives us the awesome power to make evolutionary (and scientific) leaps quickly—in the context of billions of years of evolution, what might be considered the blink of an eye.

The scientific evidence has also convinced me, a once-committed, rock-solid atheist, of the spiritual nature of the Universe. The exciting, counterintuitive findings of quantum physicists, including my first scientific hero, Albert Einstein, led me to Gnosticism, the belief that human beings contain a piece of God (the highest good or a divine spark) within themselves. Around 180 AD, Bishop Irenaeus of Lyons outlawed Gnostics as heretics and enemies of the Church.

But the stranglehold the Church once had is over. Today, Gnosticism is back, not just as a philosophy but as modern science defined by quantum physics. As physicist R. C. Henry of Johns Hopkins University wrote in *Nature*: "The Universe is immaterial—it's mental and spiritual. Live and enjoy."[1]

And I would add: live, enjoy, and *hope*. We can all emerge

from our current survival-of-the-fittest, ultracompetitive civilization to the final (and better!) level of evolution, an ultracooperative, peaceful civilization in which all humans can live and enjoy.

DEFINING CONSCIOUSNESS

Before I launch into the details of my theory, I need to explain why I call it *conscious evolution* by first defining what I mean by the word *consciousness*. Traditionally, consciousness has been thought to be a metaphysical process outside the realm of conventional science, but as a cell biologist I believe that understanding is far too narrow. Human consciousness is indeed a special and spectacularly advanced version of consciousness, but human consciousness didn't suddenly and randomly appear! Instead, consciousness evolved along with the evolution of life. To better understand that wondrous trajectory, in this book I focus on a broader and more basic definition of consciousness:

> *Being aware of, and responding to, our surroundings, the environment.*

It was American evolutionary biologist Lynn Margulis who first challenged conventional beliefs about consciousness when her work established that bacteria display consciousness, albeit a more primitive version of it.[2] When microscopic proteins, the building blocks of life, move in response to very specific environmental signals, I believe those proteins are displaying a most basic form of consciousness. That is why I consider integral membrane proteins (IMPs), which act as the interface between chemical and energetic environmental signals to control biological behavior, including genes, to be the superstars of conscious evolution, and why I define them as "perception units" throughout this book. (In this book you'll hear lots about IMPs and the "magical" cell membranes within which they are embedded.)

Tracking the evolution of consciousness-generating molecules led Margulis, as it has led me, to an evolutionary theory that challenges the competition-focused theory of Darwinian evolution. Before her theory became orthodoxy, Margulis faced ridicule when she determined that the eukaryotic (nucleus-containing) cell emerged from the *symbiotic* union of primitive prokaryotic cells. That insight prompted her to famously conclude: "Life did not take over the globe by combat, but by networking."[3]

As I've argued in all of my books and lectures, we "advanced" humans can learn important lessons from "primitive" cells whose evolutionary development, as I repeat over and over in this book, is that harmony, and most importantly cooperation/networking, is the best way to ensure that our teetering civilization not only survives but thrives!

Chapter 1

LAMARCK'S RED PILL

Though I have been a passionate cell biologist since the magical moment I first peered through a microscope as a seven-year-old, I had a lot to learn before I became a competent one! I fulfilled my lifelong dream of studying and teaching about cells in 1971 when I earned my doctorate degree at the University of Virginia based on my research of cloned stem cells in tissue culture. This research continued when I became a professor at the University of Wisconsin's School of Medicine. To the astonishment of my colleagues (and, I confess, to me), I resigned my Wisconsin tenured professorship and became persona non grata in the scientific establishment.

The behavior of cells I observed in the tissue cultures led

me to question what had become science's *Central Dogma*: the belief that genes control biology. Fortunately, the significance of my research was recognized by the chairmen of the Departments of Pathology and Dermatology at Stanford University's School of Medicine, who offered me a position as a senior research scientist.

What I've been up against since I rebelled against scientific orthodoxy is the same groupthink that some of my heroes in history have been up against. Whether it's Galileo taking on a Church wedded to the notion that planets revolve around Earth, much vilified Jean-Baptiste Lamarck taking on creationism with his theory of evolution (fifty years before Darwin!), or Lynn Margulis's paradigm-smashing discoveries about bacteria, scientists who blaze a new trail do not have an easy time of it.[1]

Not that I'm putting myself on par with the giants of scientific history, but they continue to inspire me. My cell-cloning research led to my heretic status, but things got worse, as evidenced by the New York editor who labeled me a creationist, when I realized that Lamarck's theory of evolution, which stressed a symbiotic and cooperative relationship between organisms and their environment—not Darwin's survival-of-the-fittest natural selection—dovetailed with my cell-cloning experiments.

With knowledge not available to the pioneers of history, I started to develop my theory of evolution. Scientists had unraveled the mystery of the structure and function of the hereditary molecule DNA (deoxyribonucleic acid), but I was learning that this Nobel Prize–winning breakthrough did not tell the whole story. In my lab, I was discovering the Lamarckian lesson that the fate of my cells depended not on their DNA-containing nuclei but on their environment, that is, the tissue-culture medium I provided in the petri dish.

Nevertheless, it was Darwin, not Lamarck, who took the nineteenth century by storm after he published *The Origin of Species by Means of Natural Selection*, prompting anthropologist Loren Eiseley to dub the century "Darwin's Century." He more accurately could have named it the "Darwin-Wallace Century" after Alfred Russel Wallace, the century's greatest explorer-naturalist, who co-discovered the theory of natural selection and "championed the theory as vigorously as did Darwin," according to biographer Ross A. Slotten in his 2004 book *The Heretic in Darwin's Court.*[2] In fact, notoriously cautious Darwin, who knew what a cataclysmic effect his theory would have on the ecclesiastical and scientific elite (of which he was a part), kept collecting data rather than publishing his theory until he was propelled by Wallace's 1858 co-discovery to finally publish *On the Origin of Species* in 1859.

Darwin went on to take the twentieth century by storm as well. My last job in academia was at Stanford University where I, as I had throughout my twenty-five-year career, cloned stem cells. One day I went to the evolution section in one of the Stanford libraries hungry for alternative theories of evolution. Instead, what I saw was row after row of heavy tomes, all spin-offs of Darwinian theory. Then a eureka moment when, on the last shelf, I found a slim, recently published book, *Beginnings of Cellular Life*, by biophysicist Harold J. Morowitz.[3]

Here was an esteemed scientist who shared my passion for cell biology and shared my belief in the importance of the "magical" cell membrane that makes interaction with the environment possible.

I was particularly taken by Morowitz's statement that "life arose on Earth by processes that may be understood by considering the laws of chemistry and physics as applied to complex *adaptive* systems." The emphasis on adaptive is mine because it echoes Lamarck's theory of evolution that I had come to admire.

At the time, you would have been hard-pressed to find another academic who called Lamarck one of his/her scientific heroes as I did (another longtime scientific hero, Albert Einstein, has always been a far less controversial choice). But in recent years, Lamarck and his theory, once sentenced to obscurity, have made a comeback thanks to the burgeoning field of epi- ("above") genetics, whose researchers are pinpointing chemical additions to DNA that enable organisms to adapt to environmental signals and pass those changes on to future generations *without changing* the structure of the DNA.

As evidence mounted for heritable epigenetic adaptions

that do not change the structure of DNA, some scientists started calling themselves "neo-Lamarckians" and even evolved from asking the question posed in 2000 in an article in the prestigious journal *Science*, "Was Lamarck Just a Little Bit Right?" to acknowledging that the most derided aspects of Lamarck's theory (heritable adaptive changes) were *quite* right![4]

They include paleontologist and astrobiologist Peter Ward, author of the 2018 book *Lamarck's Revenge: How Epigenetics Is Revolutionizing Our Understanding of Evolution's Past and Present*."[5] Ward writes that Lamarck is "one of the great tragic giants in the history of science . . . in originating and then promoting his novel ideas, Lamarck challenged the establishment of the best-known naturalists of his time and for his ideas he was hounded to the point of loss of money, reputation, and then health."

Lamarck was the first scientist to publish a comprehensive theory of evolution in his 1809 book *Philosophie Zoologique,* but he was on the losing side of a battle with another French intellectual, Georges Cuvier, the "founding father" of paleontology. Cuvier resisted the contention by early evolutionists like Lamarck that the age of the earth could be measured in millions (it's up to billions now!) not thousands of years. Instead, he stuck to the Church's belief that Earth's age coincided with biblical history. He also believed in "intelligent design" (as some creationists still do) that holds that God repopulates Earth after mass extinctions.

A wealthy and apparently intensely jealous aristocrat, Cuvier hounded Lamarck (also born into an aristocratic family but alas an impoverished one) in life and even in death when he delivered his eulogy damning Lamarck and his theory of evolution at the French Academy of Sciences in Paris. Writes Ward: "Cuvier did everything he could to intellectually assassinate Lamarck, and in fact he would have danced on his grave had not the aged, blind, and penniless Lamarck been 'buried' in a shallow pauper's pit of lime rather than a grave at all, his flesh and bones rapidly consumed."

What does this nearly two-hundred-year-old feud between Lamarck and Cuvier teach us other than to confirm that science has never been exempt from petty jealousies and that challenging scientific and religious orthodoxy carries a high price? I believe that the Darwinian view of Nature memorably summed up by his contemporary, poet Alfred Lord Tennyson, as "red in tooth and claw," has had a corrosive and largely unrecognized effect on our culture, an effect that could have been mitigated had Lamarck's much less harsh theory of *cooperative* interaction between organisms and their environment not been buried in scorn.

My audiences are hungry for my New Science–informed theory because it provides hope that they can emerge from the chaos that surrounds them even though they understand the havoc humans have wreaked on this planet's finite resources. Given how we're squandering Earth's resources, Global Footprint Network has calculated that we rapacious humans currently need 1.6 planet Earths to sustain us. That squandering includes widespread deforestation, plastic pollution, sectarian violence and war, and climate change that is not only melting the ice caps but also producing erratic weather patterns that threaten farmers' ability to feed the world's eight billion people.

During my lectures, I place our current situation into an evolutionary context by contending (as many biologists do) that we are on the threshold of a sixth mass extinction event. Darwin and others believed that evolution is a very slow and gradual process. Informed by Stephen Jay Gould and Niles Eldredge's theory of punctuated equilibrium (1972), scientists now believe that evolution is the result of long periods of stasis punctuated by sudden, dramatic upheavals, which is why my coauthor Steve Bhaerman and I titled our 2009 book *Spontaneous Evolution.*

In the last 600 million years, our planet has experienced five dramatic mass extinction events that wiped out 75–90 percent of life. The most recent event, estimated at 66 million years ago, occurred when an asteroid hit Mexico's Yucatán Peninsula, putting an end to the reign of Hollywood's favorite, the dinosaur.

The difference between the looming sixth extinction and the previous ones is that this time there are no meteors or collisions with comets or massive geological upheavals to blame—at least not yet—this time we have only ourselves to blame.

A 2015 paper published in *Science* argues that as a species, humans "function as an unsustainable 'super predator'" fourteen times more lethal than predators in the wild. When humans get involved, we *wipe out* our prey whereas predators in the wild "generally achieve a balance with their prey, populations such that they both persist."[6] Thanks to super predator humans, "Currently, the species extinction rate is estimated between 1,000 and 10,000 times higher than natural extinction rates."[7] My audiences may not have heard human beings labeled/denounced as "super predators," but I can tell from their nodding heads, worried looks, and despairing questions that they find the term apt.

A study sponsored by NASA has highlighted the prospect that global industrial civilization will likely collapse in the coming decades due to unsustainable resource exploitation and increasingly unequal wealth distribution.[8]

The most crucial contributor to a potential apocalypse is human civilization's separation from Nature. Ten thousand years ago, the Indigenous people of the world recognized Nature is a Garden and their role was that of gardeners who supported "Mother" Earth.

The spread of Judeo-Christian religions introduced Genesis, the biblical origin story. According to the Genesis narrative, the first five days of creation brought into being the Universe, Earth, and Nature. It was on the sixth day that humans were added to the mix, a scenario that led to the belief that humans are a separate creation from Nature. The

Church, claiming infallible knowledge direct from God, emphasized that humans had "dominion" over the Garden.

The field of modern science was introduced in 1548 following the death of Nicolaus Copernicus, a Polish astronomer and mathematician. Copernicus was a priest, a doctor, and an astronomer whose work established the fact that Earth is a small planet surrounding the sun. His discovery of a "heliocentric" (sun-centered) Universe proved that the so-called infallible Bible was in fact quite wrong.

Copernicus's research opened the door for science to question and challenge the Church's formerly restricted dogma. Enter Francis Bacon (1561–1626), a prominent English lawyer, philosopher, and champion of modern science. Bacon held that by understanding the material realm, humans would gain mastery over the natural environment. A devout misogynist, Bacon presided over witch trials and used the metaphors of domination and sexual violence to describe the scientific method; his imagery played a profound role in justifying the subjugation of women and the natural world.

Darwinian evolution theory served to soundly sever science from religion's restrictive dogma. However, its emphasis on heredity as the controlling force of evolution also separated human connection with Nature. Because in his theory, evolution is mediated by "random" genetic "accidents," a belief took hold that humans arose as accidents, independent of environmental "control" with no apparent responsibility in the unfolding of planetary life.

After the discovery of the double helix, science doubled down on its course to dominate Nature using the understanding of hereditary mechanisms and genetic engineering that offered an opportunity to control not just human life but also ecology. Using misappropriated scientific discoveries, industrial agriculture, the pharmaceutical industry, and petroleum, corporations have been systematically destroying the environment and precipitating the planet's sixth mass extinction . . . which of course, includes human civilization.

I believe that Darwinian theory's "survival of the fittest" is one of most destructive influences that is currently driving humanity to the precipice of extinction. The most serious aspect of this directive is that it's a "win at any cost" directive, devoid of moral constraints, and devoid of concern for the vitality of civilization or the vitality of the biosphere. Civilization's "winners" are judged in terms of having the "most toys," which in today's world means money . . . hence, Darwinian success is motivated by the greed that drives the "success" of the 1 percent.

In contrast, the profound insight offered by conscious evolution is that evolution is not based on the behavior of individuals. Instead, it is predicated on the integration and influence of the whole community and how they relate to the harmony of the biosphere.

The good news is that civilization has time to change its ways. Through energy conservation efforts and learning to live in harmony with Mother Earth, we can turn this ship around and set it on a course for sustainability.

In *Lamarck's Revenge*, Ward uses seventeenth-century London as an example of how cultures can change quickly. Between Black Plague outbreaks, there was a golden age of Elizabethan theater, including the heyday of Shakespeare and his Globe Theatre. But only two short decades later, the theaters were shuttered by Puritans, the religious fundamentalists whose reign included the execution of King Charles I and the arrival of despot Oliver Cromwell: "In a very short period of time, the mores of an entire society shifted, and the Puritan ideal allowed no drinking, dancing, and on and on."

Because I'm an optimist, I like to focus on the Elizabethan theater portion of Ward's example. And I also like to offer another example of speedy change that I call the "Honeymoon Effect" (and wrote a book about). Everything can look overwhelmingly bleak, and then one day we fall in love and suddenly experience the "Honeymoon Effect"—we see everything through rose-colored glasses and find ourselves playing Louis Armstrong's "What a Wonderful World" over and over.

The challenge, as we all know, is to maintain that positive worldview when the glow of the honeymoon effect wears off. In *The Biology of Belief*, I stressed how individuals can deprogram the negative thoughts and self-destructive myths lodged in our subconscious minds using energy psychology.

In this book I stress that we can do the same thing as a culture.

The iconic movie *The Matrix*, it turns out, isn't science fiction, it's nonfiction. We as a culture have been programmed with four self-destructive myths, which I write about in this book and stress again in chapter 11 as surely as the Church inculcated the belief that Earth is the center of the Universe and Gnosticism was heresy. Once we let those myths go, once we swallow the proverbial red pill, we can take the next evolutionary step: creating a new civilization focused on cooperation, love, and harmony.

Chapter 2

INSIGHTS FROM A FRACTAL UNIVERSE

I'm starting with fractal geometry, because its patterns underpin my optimism about where evolution can take us. A fundamental characteristic of fractals (the word was coined in 1975 by IBM mathematician Benoît Mandelbrot) is that their structure reveals a hierarchy of nested "self-similar" patterns from their lowest to highest scale. The eternally popular, hand-painted Russian nesting (Matryoshka) dolls provide a rough idea of the nature of nested, repetitive, self-similar images. Each smaller version of the doll is a miniature, though not necessarily exact, version of the larger doll in which it is nested.

Matryoshka dolls, also called Russian dolls, are a set of brightly painted hollow wooden dolls of varying sizes, each designed to nest inside one another.

Fractal geometry's use of a simple, repetitive equation results in an iteration, a repetition, of self-similar shapes in the physical world. When you view nature through a fractal lens, iterated patterns offer predictability in structure and behavior. For example, the structure and function of a cell is self-similar to that of a human, and the behavioral patterns of animal communities are self-similar to the behavioral patterns of human communities. Those iterated patterns led me to the following analysis of the evolution of systems and vertebrate species, the first of several diagrams I offer in this book to help explain my theory of conscious evolution.

THE EVOLUTION OF SYSTEMS

One example of a pattern that reoccurs throughout the biosphere is the evolution of systems. A *system* is a set of interdependent parts that collectively interact to form a complex, unified, whole organization designed to achieve a specific purpose. Systems can be physical, like a biological system or solar system, or conceptual, e.g., a system of laws or medicine. Some systems display both physical and nonphysical components.

Examples of different forms of systems include *physical systems* like the respiratory system; *conceptual systems*, like a political system; *social systems*, like an educational system; and *information systems*, like computer networks. A property of all systems is that their function or purpose expresses an emergent character, i.e., the sum of the whole is greater than the sum of their individual parts.

Fundamentally, all systems can be defined as having the following interactive functional components organized into a complex whole: *input*, *process*, *memory*, and *output*. Conventionally, to understand a system, we focus on studying the nature of each of the component parts, and with that awareness, then try to integrate their activities into a collective whole.

Panel **A** of the figure below reveals the relationship among the basic functional components of a complete system. Inputs and outputs are the system's interface with the environment. Integrating the input stimuli with a functional output is the role of process. In carrying out its behavior-generating activity, the process function interacts with previously recorded memory, downloaded information derived from previous experiences that are included in the processor's calculations. The result of all the processor's calculations produces a response that is expressed as the system's output. Systems did not arise in their final, completely functional complexes but instead evolved over time and in discrete stages.

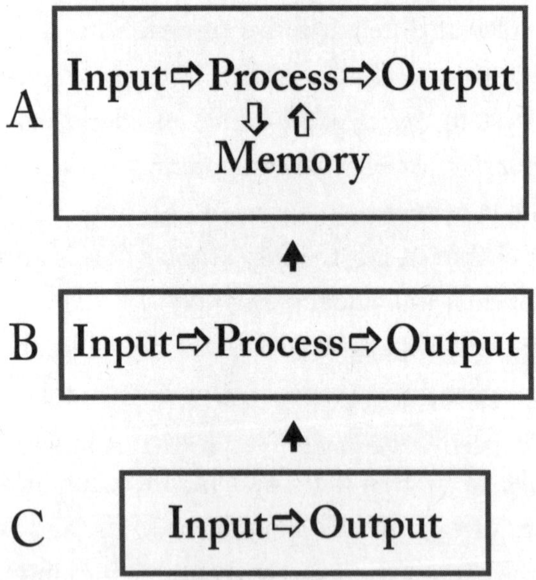

Sequential stages in the evolution of a system.

A question arises as to which of a system's components can be eliminated and still end up with a functional output. As illustrated in panel **B**, removing the memory component still allows inputs to be processed, which would lead to a functional output. Panel **C** reveals an even simpler form of an assembly that can still provide a response to incoming information. In its most basic form, inputs can directly activate specific outputs. This would be tantamount to "reflex" behavior.

The evolution of each complete system included three stages of development: **C** > **B** > **A**. The most primitive organization, illustrated in **C**, represents simple *reflex behavior* (In > Out) mechanism. The more advanced organization in **B**, which introduces a **process** mechanism, provides for a more complex response in connecting inputs with outputs. Lastly, the addition of **memory** completes the assembly of a whole system (illustrated in **A**), by incorporating learning experiences in calculating a response to incoming signals.

To better understand how any system works, it is best to first study thc earliest evolutionary form of a system and then trace the system's evolution to its more advanced versions.

In cellular evolution, the most primitive form of behavior is associated with single proteins that simultaneously expressed the functions of both receptors (inputs) and effectors (outputs). This composite protein would express simple reflex behavior as in Figure **C** on the previous page: input direct to output.

With evolutionary advancement, the function of protein-receptors were separated from the action of effector proteins, as illustrated in Figure **B** on page 32. Separating receptor and effector activities into different protein families allowed for multiplexing, wherein one environmental signal could simultaneously activate a number of different outputs, or a single behavioral output could be activated by a variety of receptor inputs. A family of processor proteins, for example, *G-proteins*, evolved to mediate the connection between incoming environmental signals and the activation of cell behavior.

A human body is an example of impressive evolutionary advancement because it is a complex system comprised of a variety of integrated subsystems, including the nervous system, the digestive system, and the respiratory system.

Assessing the evolutionary stages of systems, including the human body, offers important insight into understanding the process and mechanics of biological evolution. Reverse engineering the evolution of the human body offers a guide to sequencing the stages that led to the creation of biological organisms, from cells to humans.

Accordingly, the first level of biological evolution is comprised of input/output mechanisms, the proteins that provide for information-receiving receptors and behavior-producing effector proteins. Evolution followed with the creation of processor molecules, including connector proteins and RNA molecules that collectively enable multiplexing between incoming environmental signals and outgoing behaviors. Lastly, the evolution of RNA and DNA molecules completed the evolution of a biological system by providing for the memory of learned behaviors.

THE EVOLUTION OF VERTEBRATES

Vertebrate species evolved through a patterned sequence of increasingly complex stages that led from fish to amphibians, to reptiles, birds, and finally to mammals. In a fractal universe, the evolution of human communities would develop in self-similar stages. Let us compare.

Vertebrate Evolution

Fish	Amphibians	Reptiles	Birds	Mammals
Mariculture	Agriculture	Industrial Age	Aviation	Nurturers

Civilization Evolution

Chart comparing the evolution stages of vertebrate organisms with the evolution of human civilization.

Fish: The fundamental characteristic of fish is their dependence on a water environment. The earliest self-sustaining human communities were, like fish, physically restricted to the immediate vicinity of water. Such human *mariculture* societies flourished by cultivating and harvesting the seas and lakes and spread their civilizations across the globe by sailing from one coast to another.

Amphibians: Though amphibians are born in water, by acquiring mechanisms to take water with them, they are able to venture onto land. Human civilization entered its amphibian phase when people were able to leave their mariculture environment behind. This phase was marked by the rise of agriculture that enabled humans to sustain themselves on land, thanks to their development of methods to move water to, or extract water from, arid land.

Reptiles: Amphibians, relatively sluggish and vulnerable on land, lost their dominance in this realm with the appearance of reptiles. Reptiles left their amphibian aquatic skills behind when they perfected a superior, land-based physiology. The machinelike adaptations of reptiles evolved into bodies with great strength, speed, and dexterity. The digital characteristic of lizards' darting eyes, tongue, and mechanized gait attests to their machinelike nature. Human civilization followed a self-similar path when the Industrial Revolution led to humanity's transition from its earlier agricultural (amphibian) phase to a more sophisticated machine (reptilian) age.

Dinosaurs: This unique branch of highly successful reptiles started out as small organisms and eventually evolved into giant "killing machines" that ruled the world. Nature essentially used the blueprint of a five-inch lizard to model the formation of a fifty-foot *dinosaur*, a Greek word meaning "monstrous lizard."

While these monstrous lizards' bodies grew to massive size, their brains remained small. A proportional growth in brain size wasn't necessary for the following reason: Consider that it takes ten muscle cells to move a lizard's leg through a specific motion and only one nerve to control that behavior. The massive dinosaur leg is so large that it took ten thousand muscle cells to create that motion but still required only *one* nerve to activate it. While dinosaur bodies became massive, their brains remained quite small. The fact that lizards are still here and dinosaurs extinct suggests that the dinosaur's undersized brain, though able to support amazing reflex behaviors, wasn't adequate to sustain the survival of their massive bodies in environmental upheavals.

Human civilization is at a similar stage. The successes of the industrial age enabled human industries to evolve from small mom-and-pop shops into giant, ruthless international corporations as well as governments, organized religions, healthcare providers, and academic institutions. Similar to the pattern of their dinosaur predecessors, these behemoths are characterized by large bodies (i.e., administration and workers) and small brains (executives). Like dinosaurs, their "brains" control the reflexive behavior and growth of their organizations . . . as long as the environment stays stable. But dinosaurs and today's giant organizations lack the neurologic ability to control and adapt their massive "bodies" to survive environments in upheaval.

Another interesting similarity between today's giant entities and the former real-life dinosaurs is that our culture is fueled by oil, the "blood" of the dinosaurs. As industrial civilization drains the last of the planet's "blood," our reptilian version of civilization is facing its own imminent extinction, or at least the demise of its ruling dinosaurian organizations.

Birds: Though dinosaurs were the first animals to "rule" the world, two other evolutionary lineages, birds and mammals, were rising in the shadows. The evolution of birds was a direct spin-off from their dinosaur precursors. A self-similar pattern of evolution in human history occurred when organizational "dinosaurs" paved the path to humanity's bird phase, an event marked by the 1903 flight of the Wright brothers over the sands at Kitty Hawk, North Carolina.

The introduction of the aviation age radically altered the planetary development of human civilization. Before aviation, Earth's massive size was a formidable barrier to integrating the world's population. However, after aviation the world got much smaller. Today, given jumbo jets and the internet, physical distance is no longer a relevant factor in either peace or war.

The pinnacle of humanity's bird phase occurred in 1969

when Apollo astronauts Neil Armstrong and Buzz Aldrin stood on the surface of the moon and beamed back pictures of Earth. When the astronauts' view of Earth as a blue-green gem suspended in the black emptiness of space was sent back to our planet, this single image had a profound effect on civilization and the course of evolution. The image of the earth suspended above the moon's horizon instilled a sense of the planet's fragility, which propelled an ever-increasing number of former "reptilian" phase humans to a leap in consciousness; they became aware that our survival depends on *nurturing* the planet and ourselves. Less than a year after the Apollo photo was seen by the public, consciously awakened individuals created the first Earth Day and the seeds of the planet's next evolutionary leap, human civilization's mammal phase.

Mammals: While birds were evolving from the dinosaur lineage, a new species had split off from the original line of small reptiles. This furry species was a new class of neurologically more sophisticated (more conscious) organisms: mammals. Because of the way they raise their young, mammal species are characterized as *nurturers* who encourage growth, development, and thriving.

A massive comet crashed in the Yucatán Peninsula 66 million years ago resulting in a mass extinction that included all the dinosaurs. Up to then, small reptiles and meek mammals were at the mercy of the "monstrous lizards." When dinosaur "killing machines" disappeared, mammals had an opportunity to evolve and become masters of the biosphere.

The current ecological, climatic, and economic crises precipitating today's global chaos are an iteration of a self-similar pattern that occurred when dinosaurs and primitive, furry mammals shared an uneasy coexistence on the planet. They are also portents of the impending demise of reptilian "dinosaur" organizations and communities and the rise of Earth-friendly humans, our "mammal" phase when we nurture others (and ourselves) and our planet rather than seeking dominion over them.

Currently, civilization's chaos is a sign that we are preparing to make the ultimate, giant evolutionary leap, the creation of a superior new organism . . . Humanity with a capital *H* that I'll use throughout this book to accentuate the fact that humanity hasn't yet made the ultimate evolutionary leap to superorganism Humanity.

Chapter 3

CONSCIOUS EVOLUTION AND THE TREE OF LIFE

In his 1879 book *The Evolution of Man*, zoologist Ernst Haeckel's Tree of Life illustration traced evolution from simple, single-celled organisms at the bottom of the trunk to humans at the top, as illustrated on the next page.

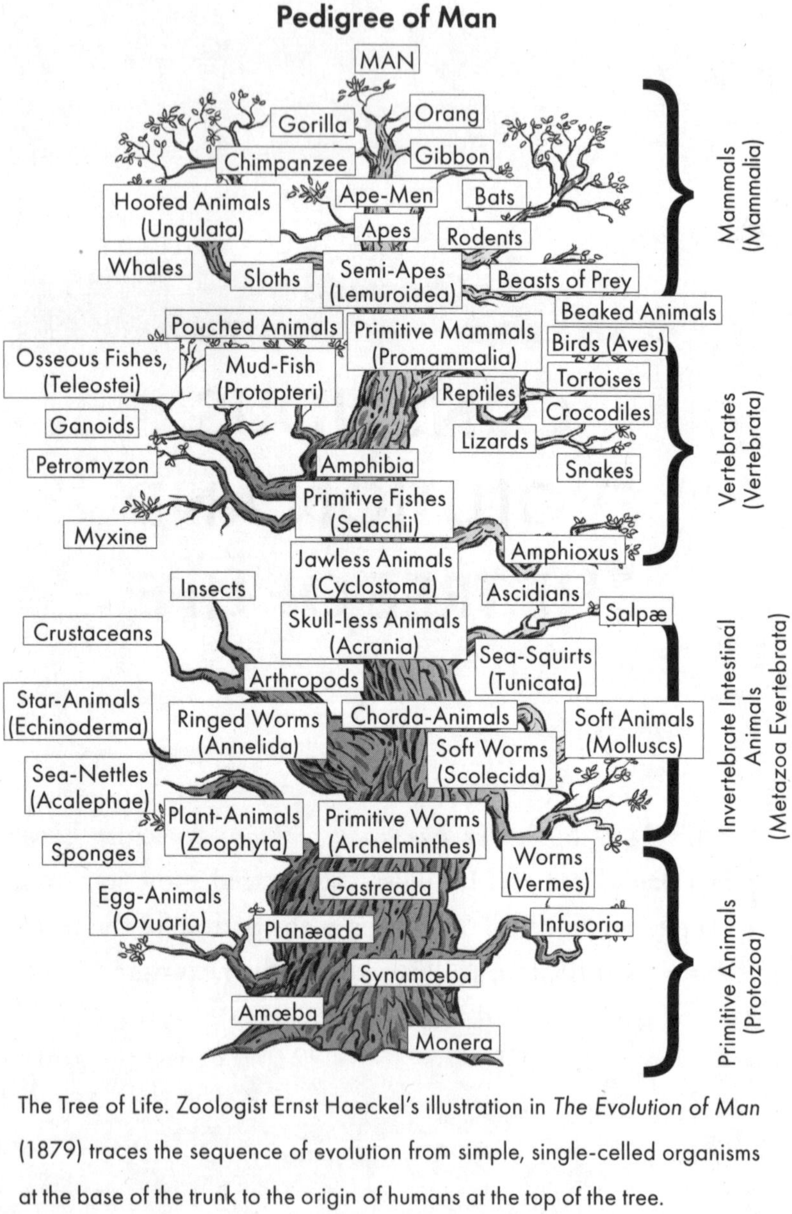

The Tree of Life. Zoologist Ernst Haeckel's illustration in *The Evolution of Man* (1879) traces the sequence of evolution from simple, single-celled organisms at the base of the trunk to the origin of humans at the top of the tree.

When scientists embraced the Central Dogma, the belief that genes control life, the assumption was that the higher up the tree, the greater the genetic complexity. That assumption stemmed from James Watson and Francis Crick's 1953 Nobel Prize–winning discovery that the structure of the DNA molecule is a "double helix," two polynucleotide chains coiled around each other, a twisted ladder that carries genetic instructions for the development, functioning, growth, and reproduction of all known organisms and many viruses. (Actually, in another Lamarck-like travesty of scientific history, it was the "forgotten heroine" Rosalind Franklin whose research first produced X-ray diffraction images of the double helix, images that her major advisor passed on to Watson and Crick without her knowledge or permission, and for which, unlike Watson and Crick, she received only posthumous credit and no Nobel Prize.)

Given the fact that it takes a gene "blueprint" to synthesize a protein molecule, and the human body contains in excess of 100,000 different proteins, it was expected that the 1990–2003 Human Genome Project would map more than 100,000 different genes. However, scientists were caught with their genes down when it was discovered that highly advanced humans had 22,000 genes, the same number of genes as found in lowly worms. Those astonishing results prompted science writer and environmental activist Connie Barlow, editor of *From Gaia to Selfish Genes* to exclaim: "My pet salamander has 20 times more DNA than I have."[1]

On a more significant level, the Human Genome Project undermined the central dogma, biology's sacrosanct belief in genetic determinism. That dogma emphasized that the flow of information in biology is unidirectional, from DNA to RNA and then to protein. We are the "protein" because proteins are the macromolecules that make up our bodies. The point of the dogma is that we, the "protein," cannot reverse the information flow and go back and alter our genetics. This is the origin of the belief in *genetic determinism*, the notion that our lives are controlled by our genes, and that we are the "victims" of our genome. Epigenetic research now reveals that information flows in *both* directions *and* that environmental signals are the primary source of information operating above ("epi") the DNA.

In addition, it is now understood that RNA includes a large family of information molecules that collectively engage to translate the genetic code into the synthesis of each cell's proteins. The variety of RNA molecules include microRNA (miRNA), ribosomal RNA (rRNA), transfer RNA (tRNA), and messenger RNA (mRNA). The information encoded in the messenger RNA (mRNA) is a replication of a DNA code for the synthesis of a specific protein molecule. The mRNA molecule carries the copy of the DNA code from

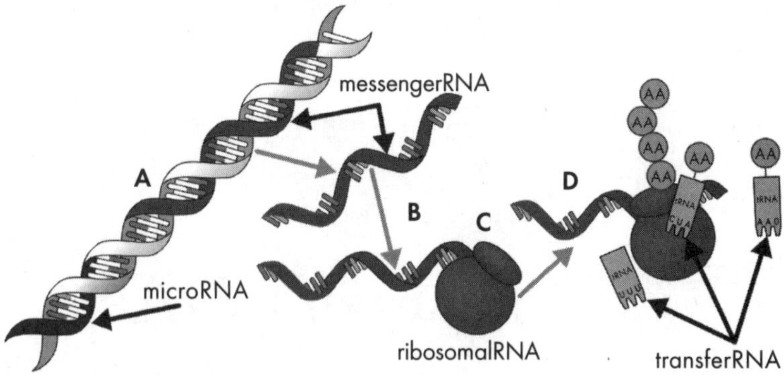

Illustration of different RNA molecule functions: (**A**) DNA strand in nucleus. MicroRNA binds to gene blueprint and regulates gene activity. *Messenger-RNA* (*mRNA*) copies gene blueprint (*transcription*). (**B**) *mRNA* moves from nucleus into cytoplasm. (**C**) *mRNA* binds to ribosomes (*ribosomalRNA, mRNA*). (**D**) When *mRNA* moves through the ribosome, transferRNA (*tRNA*) molecules carrying specific amino acids, bind to the matching *mRNA* code. The attached amino acids bond together, forming a chain that becomes the protein—this process is called *translation*.

the cell's nucleus to the ribosome (rRNA), a cytoplasmic organelle that translates the code into the specified protein.

Long before those research breakthroughs, in the late 1880s, German biologist Walther Flemming became the first to observe and systematically describe the behavior of chromosomes, the threadlike structures inside the cell's nucleus. Scientists eventually determined that chromosomes are made up of 50 percent DNA and 50 percent protein, but in the aftermath of the double helix discovery era, research

DNA — **Long Term Memory**

↓

RNA — **Short Term Memory**

↓

Protein — **Behavior**

Primacy of DNA

Environmental Signals — **Primary "Cause"**

↓

Regulatory Proteins — **Epigenetic Control**

↓

DNA — **Long Term Memory**

↓↑

RNA — **Short Term Memory**

↓↑

Protein — **Behavior**

Primacy of the Environment

This illustration on the left illustrates the conventional flow of information in biology, represented as the "Primacy of DNA," a system wherein DNA "runs" the show. On the right is the latest science on information flow. The control of biology is influenced by the environmental signals that activate an information flow that ultimately determines genetic expression and organismal behavior. Since the brain interfaces the environmental signals with our biology, it is our consciousness that actually controls the character of our lives.

focused on the structure and function of the DNA and ignored the potential function of the chromosome's proteins.[2]

More recent research reveals the vital role these proteins play in regulating gene activity and shaping the character of our lives:

1. Chromosomal proteins form a sleeve that covers, packages, and compacts linear DNA molecules, each of which can be several centimeters in length.
2. Chromosomal protein complexes called *modules* are distributed between adjacent genes along the length of the DNA. These modules are found in the non-protein coding regions of DNA formerly referred as "junk DNA."

Chromosomal modules are the sites of epigenetic control because they interface environmental signals with the control of gene programs. Some protein modules determine which genes are to be read. Other modules edit the gene code by modifying the readout of the mRNA's message, cutting and splicing (reassembling) the molecule to produce hundreds of different proteins from the same DNA blueprint. The amazing versatility of a single gene program encoded in an mRNA copy stems from those epigenetic mechanisms, which can cause a normal gene's mRNA to produce "mutant" disease-causing proteins and can also cause a mutant gene's mRNA to read as a normal gene.

Chromosome modules also tailor mRNA transcripts (copies of the DNA code), enabling proteins to be adapted, i.e., enabling the creation of a variety of different functional structures within the same organism or in other species. For example, in humans, the same keratin protein gene can be modified to create hair, fingernails, and skin. In other species, the same keratin protein is used to create claws, hooves, horns, scales, and feathers.

In traditional Darwinian theory, mutations were thought to result from *random* alterations of the genetic code that occur during cell division when replicating DNA is most susceptible to copying errors. That assumption suggested that evolution is driven strictly by chance and hence is purposeless or simply "accidental." Studying bacteria provides a good test of the random alterations hypothesis because they replicate rapidly, offering the opportunity for a high rate of mutations as well as the ability to quickly adapt to new environmental conditions through the accumulation of *beneficial mutations*.

In 1988, John Cairns's seminal (and still controversial) research on the mechanism of bacterial adaptation challenged the belief that mutations are random.[3] Cairns worked with *Eschericia coli* (*E. coli*) bacteria that had a defective (mutated) gene for the enzyme lactase, a protein that digests lactose into energy and usable metabolic by-products. Cairns inoculated the lactase-deficient bacteria into culture dishes whose growth medium was solely lactose. Since the bacteria's lactase enzyme was defective, these cells could not digest the available source of nutrients, without which they were starved and stressed. Under this nonlethal selection pressure, lactose-utilizing bacterial colonies soon developed in each culture and began to flourish.

Cairns argued that the gene alterations that repaired the defective lactase gene were not random copying errors during replication but instead were *adaptive* mutations that enabled them to survive in their "stressed" environment. In fact, the cells mutated their defective lactase genes *before* the bacteria could replicate their DNA because, in the absence of a usable food source, they were unable to divide. Cairns also emphasized that the organisms and their environment engaged in a "positive" interaction, leading to selective mutations necessary for rewriting the defective lactase gene. He pointed to the fact that the bacterial mutations were not randomly distributed throughout the genome but were instead focused specifically on the genes coding for lactose metabolism.

Other studies have reported examples of adaptive mutation or related phenomena in bacteria and simple yeast cells. Researchers have concluded that bacteria adapt to environmental changes with temporary and permanent modifications. Generally, bacteria undergo permanent adaptive evolution when the selection pressure is strong and persistent. When the selection pressures are short term, the bacteria engage in temporary adaptation.[4] In 2008, researchers at University College London concluded: "Not only do organisms show great resilience to external challenges, but emerging data suggest that they also exploit these challenges to fuel *phenotypic variation* [italics added]." This is an evolutionary innovation that enables organisms to alter their observable characteristics to accommodate short-term stresses.[5]

A 2022 study summarized some of the reported mechanisms of bacterial adaptive evolution and the methods for applying bacterial adaptive evolution in research and industrial development, for example, creating bacteria that can eat pollutant oil spills in the ocean.[6]

Research by Indiana University professor of biology and biochemistry Patricia L. Foster focuses on mechanisms that produce adaptive mutations in another strain of bacteria, *Escherichia coli*, FC40.[7] And a similar example of adaptive mutations has been found in studies of the SARS-CoV-2 viral genome during the recent Covid-driven pandemic.[8]

From an evolutionary standpoint, the most advanced example of adaptive mutations is the human immune system. Immunologically active T cells bind to and destroy invasive or altered cells while B cells produce antibodies that evolve new and very specific proteins in response to bacterial, viral, and parasitic infections.

For example, within three days of vaccination with a foreign *antigen* (an infectant agent), the immune system will create perfectly complementary proteins referred to as *antibodies* in B cells, or *antibody-like* membrane protein receptors in T cells. Antibodies are protein molecules that are precise physical complements of antigens that recognize and bond with foreign antigens, marking them for destruction. The creation of these specifically designed proteins is not a random evolutionary event but an adaptation that is entirely predictable.

In 1967, adaptive mutation discoveries were still to come so, echoing what I had been taught, I was teaching medical students traditional *genetic determinism*: the belief that genes "control" an organism's fate. However, in my lab, cloned stem cells were telling a different story. My research revealed that genes are not "self-actualizing"; they cannot turn themselves on and off. Instead, I found that genetic activation is regulated by environmental signals.[9] My colleagues, immersed in conventional genetic research, ignored these studies. Beginning in 1988, my research at Stanford University's medical school on cloned microvascular endothelial cells extended the findings on the mechanism by which environmental signals regulate gene activity.[10]

This time my research was beginning to be vindicated . . . slowly! . . . because conventional scientists had begun to recognize the New Science of epigenetics. Two years before my Stanford research was published in 1992, H. F. Nijhout offered the following simple definition of epigenetics (I have added the italics):[11]

When a gene product is needed,
a signal from its environment,
not an emergent property of the gene itself,
activates expression of that gene.

That definition can be further simplified by combining the two italicized phrases:

> ***A signal from the environment***
> ***activates expression of that gene.***

This definition of epigenetics is profoundly important because it emphasizes that genes cannot "turn on and off," an insight that eliminates the victimology associated with the belief in genetic determinism.

With Nijhout's definition and the debunked notion that more genes mean more evolutionary complexity in mind, I propose another way of looking at Haeckel's Tree of Life, one also posited by Lamarck, who not only stressed the environment's role in evolution but also hypothesized that evolution was based on the development of the nervous system.

From a Lamarckian point of view, the Tree of Life can be seen as a pictorial illustration that traces the lineage of the nervous system's evolution from primitive bacteria to human beings; or in the language of my theory, a pictorial illustration of conscious evolution, the wondrous ascent from the primitive reflexive consciousness of bacteria to the powerful, creative self-consciousness of human beings, each of whom is made up of a remarkable collection of fifty trillion cells.

Chapter 4

CONSCIOUS EVOLUTION'S TWO-PHASE SPIRAL

At a cellular level, conscious evolution was made possible by "superstar" integral membrane proteins (IMPs) that are embedded in the cell's "magical" membrane because they read and respond to environmental signals.

Specifically, a *receptor* IMP "reads" an environmental signal and provides the cell with an "awareness of the environment." When engaged by an activated receptor

IMP, a complementary *effector* IMP sends a physical signal (sensation) into the cell to engage specific, life-sustaining cellular functions, such as cell motility; the secretion of protective enzymes; and phagocytosis, the process of ingesting elements in the environment, including bacteria and parasites. Thus, receptor-effector IMP complexes are the fundamental building blocks of consciousness that provide the cell with "awareness of the elements of environment through physical sensation," as per the dictionary definition of perception.

That means the evolution of consciousness is directly related to the population of cellular receptor-effector IMPs, aka *perception units*, built into an organism's membrane. Instead of the once-prevalent belief that more genes equal more evolutionary complexity, conscious evolution necessitates a new equation: More IMPs equal more evolutionary complexity.

That insight combined with the self-similar patterns of fractal geometry led me to create the following illustration of conscious evolution as an ascending spiral, a repetition (iteration) of a two-phase process that maximized perception units, aka IMPs in organisms from bacteria to humanity.

CONSCIOUS EVOLUTION'S TWO-PHASE SPIRAL

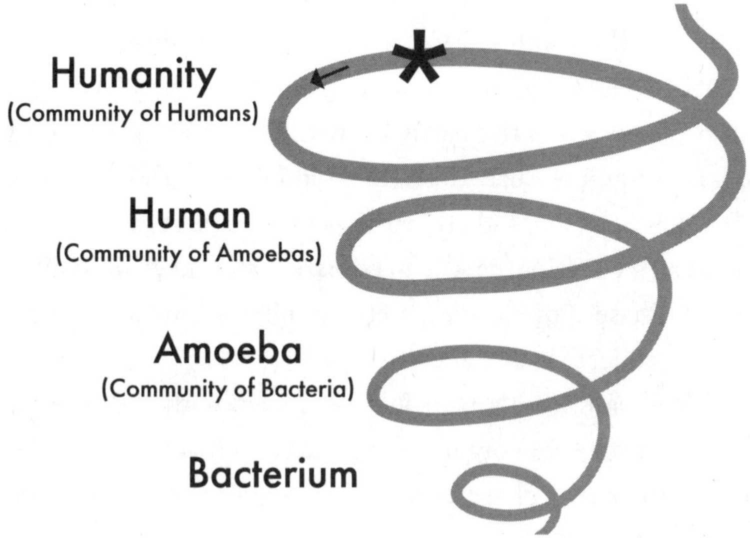

A spiral image revealing the sequential stages of communal evolution from bacteria to human civilization.

In Phase 1, the single organism evolves into its most conscious state by maximizing its membrane population of IMPs. In the process, physical constraints on the expansion of the organism's membrane surface area inevitably limit the number of IMPs an organism can manage. Evolution stops when the organism can no longer add more membrane surface area.

In Phase 2, after the most conscious single organism has evolved, conscious evolution continues when highly conscious individuals band together in community to "share" awareness and collectively increase the total number of perception units shared by that community.

After the community's collective consciousness is maximized, evolution of that communal organism stops. The cycle starts again when the newly created, more highly advanced organism enters Phase 1 on its evolutionary path to become the "smartest" individual of the new species. After that individual's consciousness is maximized, the "new" organism enters Phase 2 by forming a community of individuals who share a collective awareness.

To illustrate this principle, let's start with the Tree of Life's bottom-rung organism, prokaryotes (bacteria) and describe their two-phase evolution that led to the next, more advanced, new organism, a eukaryotic cell (e.g., amoebas) and subsequently when communities of amoebas created human beings, the highest-level eukaryote on the tree.

PROKARYOTES PHASE 1

Bacterial cells are physically supported and constrained by a rigid external "capsule," an exoskeletal support equivalent to invertebrate life-forms. Consequently, the bacterial cell membrane's surface area is confined by the limited cytoplasmic volume within its external capsule.

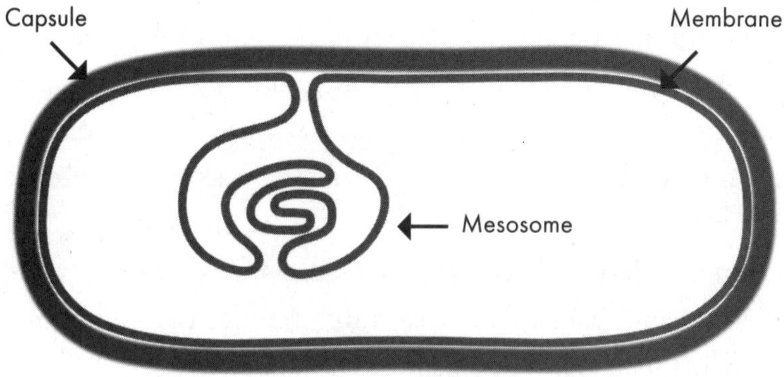

Bacterium

The cell membrane of a bacterium is constrained by a rigid exoskeleton called a *capsule*. To expand the cell's membrane surface area, the membrane forms pocket-like folds, referred to as *mesosomes*, that penetrate the cell's cytoplasm.

One method bacteria use to increase their cell-membrane surface is to create *mesosomes*, folded ("invaginated") membrane formations that burrow deep into the bacterium's cytoplasm. The bacterial chromosome contains all the genes necessary for the cell's survival. Chromosomes not only encode consciousness-providing perception units/IMPs but also encode perception units responsible for "housekeeping" chores, including digestion, excretion, respiration, cell motility, and chromosome replication.

For the bacterium to enhance consciousness, it had to evolve new receptor-effector IMPs by creating *new* genes to code for these membrane-perception proteins. The creation of new perception IMPs is essential for surviving in a changing environment. In bacteria, new IMP genes are not incorporated into the existing chromosome but float free in the cell's cytoplasm as *plasmids*, tiny circles of double-stranded DNA. Plasmid genes offer bacteria additional genetic traits beyond those found in the chromosome. For example, plasmid-derived proteins provide resistance to antibiotics and the ability to digest different forms of nutrients, such as those found in oil spills.

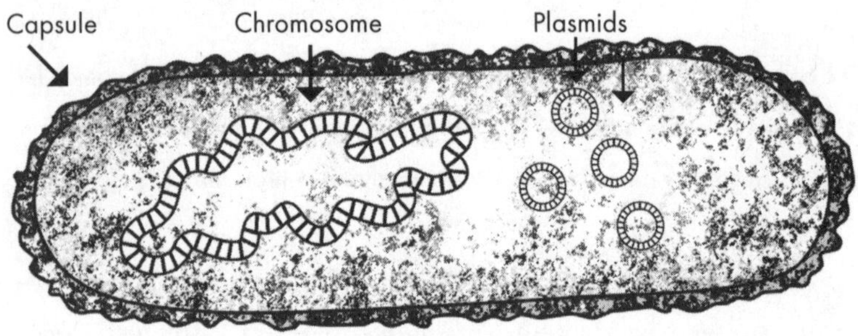

Bacterium

A *chromosome*, within the cytoplasm of a bacterium, is a large, folded DNA-protein molecule that controls the structure and function of the cell. In adapting to environmental demands, a bacterium can acquire new functions and structures that are encoded as new genes that are structurally expressed as *plasmids*, cytoplasmic rings of DNA that are separate from the cell's chromosome.

Bacteria can create an average of six to ten *different* versions of plasmids; each is derived from a unique environmentally derived *learning* "experience." Because bacteria can make hundreds to thousands of copies of any plasmids, limitations on their number is not due to an inability to make DNA. Instead, plasmid population limitations stem from the fact that each "new" plasmid-derived IMP perception unit requires additional surface area. As I mentioned above, physical constraints limit the cell's ability to expand its membrane surface area and, as a consequence, limit its ability to increase its perception, its consciousness, its awareness—all terms I use interchangeably in this book.

Extra copies of individual plasmids, the extra-chromosomal DNA rings that encode learned environmental experiences, play a significant evolutionary survival role. Consider a community of a hundred bacteria, with each bacterium containing a full repertoire of acquired plasmids and an inability to accommodate any new ones due to physical constraints.

For example, one bacterium boasts a DNA plasmid that resists penicillin. If penicillin is introduced into the environment, it could kill all the bacteria except for the one with the anti-penicillin plasmid. However, that individual bacterium can create hundreds, up to a thousand, copies of its anti-penicillin plasmid independent of copying its chromosome. In a process called *conjugation*, a bacterium physically couples with another bacterium and exchanges genetic information.

This exchange, called *horizontal gene transfer*, enables a bacterium to pass a copy of the life-saving plasmid to the other bacteria in the community, thus offering them temporary use of the anti-penicillin plasmid DNA code to save their lives.

Once the bacterium's available membrane surface area is filled with chromosomal and plasmid-derived IMPs, it has maximized its consciousness and is the "smartest," most aware individual bacterium, which marks the end of Phase 1.

PROKARYOTES PHASE 2

In Phase 2, functionally integrated communities of diverse bacterial species share their awareness to collectively enhance the community's vitality. Cooperative bacterial communities enveloped themselves in a slimy, protective matrix composed of a combination of extracellular polysaccharides (sugars), proteins, lipids, and DNA that enabled the community to survive in Earth's early harsh environment. The spontaneous assembly of a membrane around the extracellular matrix provided the bacterial community with a semipermeable barrier to protect and control the internal environment. The organization of the membrane-bound community is referred to as a *biofilm*. Examples include dental plaques, slime on shower tiles, and slippery surfaces on rocks in streams.

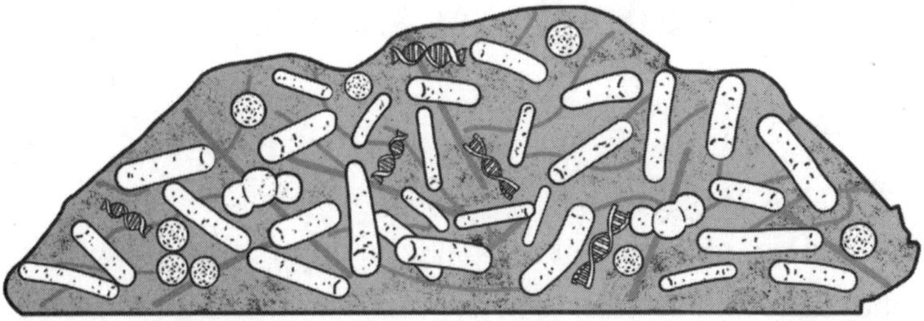

Biofilms form plaques that consist of a variety of bacterial species embedded in a slimy matrix surrounded by a semipermeable membrane. The surrounding membrane protects the community from adverse, life-threatening environmental conditions, such as the presence of toxins, including penicillin. Each bacterial species provides a specific function that collectively supports the growth and protection of the community. The extracellular matrix contains polysaccharides, extracellular DNA, and filaments such as actin that provide motility and threadlike protein fibers serving as a supporting skeleton.

That ability to regulate the community's environment enabled complex interactions that integrated communications and shared functions among the communal bacteria. Subpopulations of functionally diverse bacteria within the community cooperated in a variety of activities including metabolism, motility, the secretion of an extracellular matrix, and in adverse environments, the production of resistant spores, all processes that support the overall success of the communal lifestyle. A biofilm is the structural precursor of the next higher level of evolution, a new, nucleus-containing organism called a *eukaryote*.

EUKARYOTES PHASE 1

The evolution of the eukaryote cell was initiated by *horizontal gene* transfer, a process I introduced above. Within the biofilm, members of the bacteria population share their genome and plasmids by transferring the genes among themselves. It is probable that a specialized bacterium functioned as the community's central repository for collecting specialized genes. This proposed DNA-gathering bacterium would become the precursor of the double-membraned nucleus that characterizes an evolutionarily advanced *eukaryotic cell.*

Once the biofilm's functional genes were consolidated into the bacterial precursor of the nucleus, they collectively formed the new organism's *genome*, capable of encoding all the proteins provided by the formerly heterogeneous bacterial population. Originally, all the functional, community-derived IMPs were integrated into the enveloping cell membrane, which is referred to as the *plasmalemma* (cell membrane).

Once the primitive, large eukaryotic cell was created, it further evolved by acquiring new functional traits through a process called *endocytosis*, the engulfing of smaller bacteria. By ingesting and digesting the smaller bacteria, the larger eukaryote cell can utilize the engulfed bacterium's organic building blocks to support its own growth. However, some of the bacteria taken in by endocytosis became symbiotic cohabitants (*symbionts*) that offered new life-sustaining functions that enhanced the eukaryote cell's evolution.

Following endocytosis, up to 90 percent of the ingested symbiont's genes are incorporated into the eukaryotic host's genome.[1] There are two fundamental endocytosed symbionts that forever changed the evolutionary trajectory of life on our planet. *Cyanobacteria* were the precursor of plant chlorophyll–containing chloroplasts. A second endosymbiont, an aerobic bacterium species, presumably an *alpha-proteobacterium*, became the precursor of the cell's energy-providing mitochondria. Because the activity of these bacterial-derived organelles depended on their unique membrane functions, a small percent of original bacterial DNA was retained in the derived organelles. These are the genes required to encode the IMP-membrane specializations that generate vital functions.[2]

The original biofilm's extracellular matrix, the gel-like material that enveloped the bacterial community, contained *actin* and *myosin*, protein precursors of contractile muscle cells These muscle-like proteins provided for cell motility and the ability of the cell to change shape. Other matrix protein molecules, including *intermediate filaments* and *microtubules*, were incorporated into evolving eukaryotic cells to provide them with an internalized supporting cytoskeleton, a development that eliminated the size restrictions imposed by a rigid external capsule. As with the human body's skeleton, the eukaryotic cell's cytoskeletal proteins endow the cell with physical support and the ability to move. The internal cytoskeleton enabled the eukaryotic cell to increase its membrane surface area one thousand times greater than found in a single bacterium.

Similar to the way bacteria expanded their membrane surface area by invaginating membrane folds, or mesosomes, into the cytoplasm, the evolving eukaryotic cell invaginated and pinched off internalized plasmalemma folds, which led to the origin of the cell's membranous cytoplasmic organelles, including the *endoplasmic reticulum*, a network of membranous tubules involved in protein and lipid synthesis; the *Golgi body*, layered membranous sacs active in cellular secretion and intracellular transport; and *lysosomes*, membrane-bound vesicles containing digestive enzymes. These organellar functions were formerly specializations built into the cell's outer membrane, the plasmalemma. Once these functions were internalized as organelles, it freed up the membrane surface area and allowed the evolving eukaryote to accommodate a larger population of consciousness-generating perception IMPs in its outer membrane.

Though the internal cytoskeleton supported an increase in membrane surface area, the eukaryotic cell's size was still subject to physical constraints. As an analogy, consider filling a balloon with water. When there is a small volume of water in the balloon, it can be easily passed around from person to person. However, if too much water is in the balloon, when the balloon is passed, the pressure from the volume of internalized water will cause the balloon's skin to rupture and release its contents.

Similarly, in eukaryote evolution, there was a point at which too much internal cytoplasm in a large cell would destabilize and weaken the plasmalemma, leading it to rupture and cause the cell's death. Consequently, eukaryotic cells are subject to size restrictions, typically ranging in diameter from 10 μm to 30 μm (micrometers or microns—for comparison, the width of a human hair averages about 50 μm in diameter). As formerly experienced by the single bacterial cell, the eukaryotic cell reached its maximum size and was unable to accommodate more membrane surface area for the expansion of perception IMPs. Once the maximum IMP population was reached, Phase 1 of eukaryote evolution ceased.

EUKARYOTES PHASE 2

The evolving individual eukaryotic cells became single-celled protozoans, such as amoeba. In a pattern similar to that found in bacterial evolution phase 1, limitations on physically expanding consciousness led eukaryotic cells to assemble into multicellular communities, thereby sharing awareness and collectively raising the consciousness of the community. The evolution of amoeba-like cell communities is the basis of all the multicellular organizations we recognize as plants and animals.

HUMANS PHASE 1

Like the prokaryote cell, the size of the human brain is physically constrained inside a rigid capsule, the skull. But within that physical constraint, the human brain evolved into a marvel of nervous system development.

Vertebrate brains structurally resemble "rounded" vesicles surrounded by an *epithelium*, a multicellular surface lining that is the structural and functional equivalent of a cell membrane. The epithelial surface area of a brain, a physical reflection of its processing power, is directly determined by the size of the skull. In lower vertebrates, for example, the rat, the rounded vesicular brain conforms and fills the shape of its skull. Higher vertebrate brains, encased in larger skulls, offer the opportunity for greater processing power.

As mentioned above, primitive bacteria can increase their awareness by expanding membrane surface area through the creation of mesosomes, internalized folds of surface membranes that penetrate deep into the cell's cytoplasm. A similar mechanism for increasing brain power occurred in higher vertebrates by enfolding the brain's membranous epithelium to form layers called *sulci* and *gyri*.

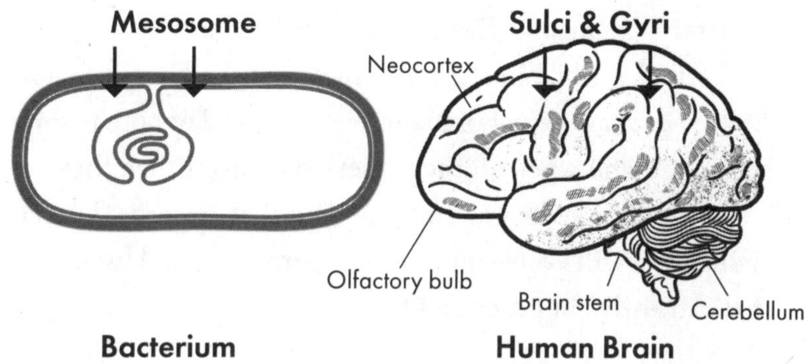

The membrane surface area of a cell is directly proportional to its processing power. Similarly, the surface area of an organism's brain is directly proportional to the intelligence of that organism. The human brain maximizes its surface area in the skull by forming folds and grooves (sulci and gyri).

When the human brain epithelium maxed the amount of folded surface area that could be contained within the skull, Phase 1 of human evolution ended with the "smartest" individual humans, though I concede that we "smart" humans do incredibly dumb things.

HUMANS PHASE 2

In humans, the Phase 2 pattern of expanding consciousness by creating communities to share their collective intelligence is still in process. Humans are learning (too slowly, I know!) that to survive, they must collectively participate as a single planetary organism. The Phase 2 evolutionary principle is perfectly described in the Latin motto of the United States, *E pluribus unum*—"out of many, one." Phase 2 of human evolution will end when all the planet's humans realize they are "cells" in the same community. When that happens, human civilization will evolve into a new superorganism, Humanity, that can thrive in peace and harmony.

Chapter 5

THE SECRET OF LIFE: TAKE TWO

Unraveling DNA's elegant double helix structure and its role in the assembly of proteins was an amazing scientific achievement, but the subsequent focus on DNA overshadowed the crucial role of proteins as the building blocks of life. The importance of the protein molecule was acknowledged as early as 1838 when Swedish chemist Jöns Jacob Berzelius coined the name "protein," wherein *pro* = primacy and *tein* = part. Though 1953 headlines touted the double helix as "the Secret of Life," to truly understand the secret of life, scientists needed to go deeper to unravel the relationship between DNA, RNA, and proteins, and most important, to understand the role quantum physics plays in those relationships.

There are over 100,000 different proteins that provide the structure and function of living organisms. Protein molecules are assembled from a family of twenty different versions of linked amino acid molecules. The linking of amino acids in protein molecules resembles a string of necklace beads. The unique character of each protein is derived from the linkage of a specific sequence of up to one thousand amino acid molecules forming a protein's "backbone." The mechanism of how life works is found in the details of the unique amino acid sequences that make up the protein's backbone.

In the family of twenty different amino acid molecules, some are "polar" hydrophilic ("water loving") molecules that dissolve in water, and others are "nonpolar" hydrophobic ("water hating") molecules that dissolve in oil. Clusters of hydrophilic or hydrophobic amino acids distributed along a protein's backbone determine whether a protein seeks stability in an aqueous or lipid environment.

Milliseconds after the protein's linear backbone is synthesized, the attractive and repulsive electrostatic forces among the amino acids in its "backbone" cause the protein to fold ("knit") into a three-dimensional *conformation* (shape) whose structure is determined by balancing the inherent electrical charges among the backbone's amino acid molecules. The process of protein folding into a stable 3D conformation can best be described as resembling "self-knitting yarn." The final stable conformation is determined by the interaction of the polar and nonpolar character of the amino acid "beads" and the influence of environmental pH and ionic electrical charges along the protein's backbone.

In its stable conformation, a protein can bond with a complementary environmental "signal" with the specificity of a lock and key. When complementary environmental signals bond with proteins, they alter the balance of inherent energy charges within the protein's backbone. When a bound environmental signal induces an energy imbalance, the peptide backbone "reknits" to form a more stable three-dimensional conformation that rebalances the molecule's new charge distribution.

As a signal-activated protein changes its conformation from its original shape to a new "active" conformation, the protein physically moves. The resulting movement represents the first step in transitioning from an inanimate world to the animate world of life. The "secret of life" at its basic level can be defined by a simple formula:

PROTEIN + SIGNAL = BEHAVIOR

Though a living person and a cadaver contain the same proteins, the difference is that in death the proteins no longer respond to environmental signals that result in movement, i.e., behavior.

Chapter 6

GENES: AN EVOLUTIONARY CONUNDRUM

While some molecular, cellular, and developmental biologists continue to unravel the intricacies of the DNA-RNA-protein building blocks of life, others focus on Origin of Life research to try to fathom the evolutionary origin of the standard genetic code (SGC) that all living organisms share. Which brings us to this chapter's chicken-or-egg conundrum: When life first emerged on Earth, which fundamental molecules, proteins or genes, appeared first?

At first glance, proteins seem like an unlikely candidate. Proteins are unstable and have a short lifespan before their amino acid backbones break down. That means in order to survive, an organism must continuously replace its degraded proteins. That is not an easy feat because the synthesis of a complex protein requires a "blueprint" for the unique sequence of amino acids in the backbone that defines each protein. How about genes? Also at first glance not likely candidates. Genes are coded sequences within the DNA molecule that are the blueprints for reproducing each protein's unique amino acid backbone. First the gene's coded DNA sequence is transcribed (copied) into an mRNA molecule. Then, as described earlier in chapter 3, a protein is synthesized when the mRNA blueprint is translated into a peptide by the cytoplasmic organelle, the ribosome.

However, one characteristic of the nucleic acid (DNA/RNA) code and its assigned amino acid sequences that has intriguing possibilities for evolutionary development is the fact that they are not directly connected. That means that for genes to evolve, there had to have been some *physical* basis for coordinating the specificity of the chemical interaction between the nucleic acid's code and the protein's specific amino acid sequence.

It is these interactions, especially the fact that individual RNA nucleotides can spontaneously bind to amino acids, that has led a number of researchers to believe they are key to understanding the evolution of the genetic code.[1]

In 2009, molecular biologist Michael Yaris and his University of Colorado Boulder team conducted an extensive mapping of the physical and energetic binding sites between specific amino acids and nucleic (DNA/RNA) molecules. In the process, they uncovered unique and high-affinity binding sites between the RNA code and each of the twenty major amino acids.[2] They also found that RNA single nucleotide bases (the cytosine, guanine, adenine, and uridine molecules), base pairs, and base triples can bind to molecules which confirmed the fact that the nucleic acid code and its assigned amino acid molecules are not directly connected.[3]

But again, which came first? Would a random sequence of nucleic acid bases generate a functional protein, or would it be more likely that a functional protein, once created, would encode a nonrandom nucleic acid blueprint?

Yarus suggests that the first step in creating the genetic code was likely functional peptides (short chains of amino acids) because the amino acids in their backbone would bind codons (any triplet combination of the four nucleic acid bases, adenine, guanine, cytosine, and uracil) at unexpected, improbably high frequencies. It is those binding capabilities at high frequencies that would have enabled them to serve as generators spurring the evolution of the genetic code. The biochemical efficiency of self-forming peptides to initiate the binding of nucleotide sequences supports an evolutionary sequence in which proteins precede the origin of RNA and DNA genetic codes.

Chapter 7

THE MAGICAL MEMBRANE

And now my favorite topic, the "magical" membrane. I've already explained in the two phases of evolution chapter (chapter 4) that the number of integral membrane proteins (IMPs) embedded in the membranes of organisms is an indicator of their level of consciousness, which is why I refer to them as "perception units." In this chapter, I'll explain why the self-assembly of biological membranes was a major event in the evolution of life on Earth, how conscious evolution can be modeled mathematically, and, finally, the important implications of the fact that the cell membrane is the functional equivalent (homologue) of a computer chip.

Evolution biologists hypothesize that membrane phospholipids, like other organic compounds, were spontaneously created on pre-life Earth when the planet's cooling crust (magma) was covered by a warm ocean. Meteor impacts and volcanic activity caused Earth's magma to release chemical and gaseous matter into ocean water. Among those chemicals were *alkanes*, long-chain oily hydrocarbon polymers comprised of linked sequences of methylene (CH_2) subunits, a fundamental molecular ingredient found in the family of organic molecules.

Through the action of solar UV, lightning storms, heat from lava, and natural radioactivity, water enriched with phosphates, nitrogen, carbon dioxide, and minerals spontaneously created phospholipid molecules. Because they are low density, phospholipids float to the surface in water environments. To stabilize the alkane's hydrophilic-hydrophobic (polar-nonpolar) *amphiphilic* regions in a water environment, phospholipids spontaneously formed a membrane bilayer with polar exterior surfaces and a nonpolar interior.

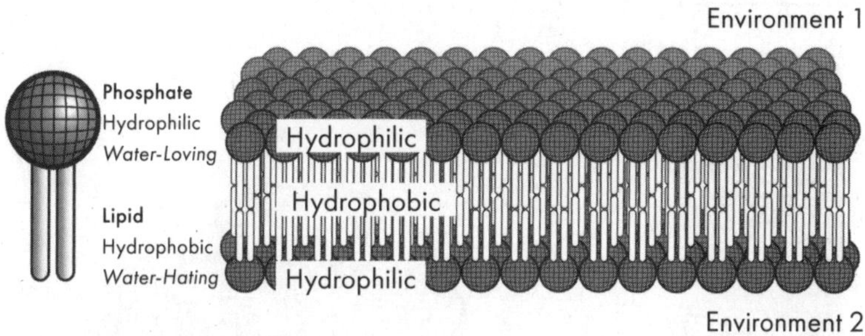

Phospholipid **Bilayer Membrane**

Lollipop-shaped phospholipid molecules arranged like at-attention soldiers in perfect alignment are the major component of all bio membranes. The physical characteristics of amphiphilic phospholipid molecules that include hydrophilic (water-loving) and hydrophobic (water-repelling) regions enable them to self-assemble into membranes.

The presumed existence of a "primordial soup" of organic compounds would not have provided the stability needed for the evolution of living organisms because the management of protein behavior requires a controlled environment (e.g., pH, salt balance, and temperature). The controlled environmental domain provided by membranes was a crucial evolutionary step in creating life because of its ability to regulate conditions required for metabolic behavior.

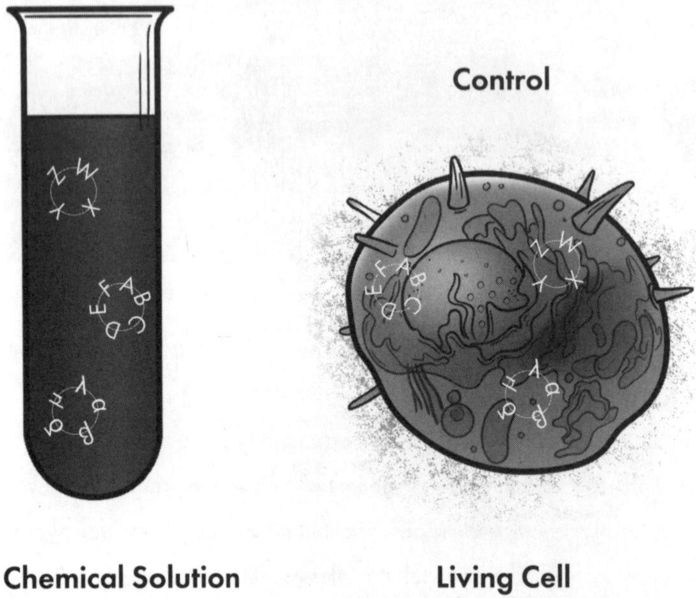

Comparing the functional difference between a primordial soup of proteins and the proteins in a cell. Proteins in a cell soup are controlled by the vagaries of an ever-changing environment. In contrast, proteins in the cytoplasm of a cell are controlled by the intelligent processing offered by the cell membrane, independent of environmental conditions.

And membranes do more than provide a controlled environment. By reverse engineering cell metabolism, biophysicist Harold Morowitz, whose book inspired me decades ago, found that in addition to controlling a cell's environment, membranes also provide an energy source to power life's processes.[1]

Phospholipids found in space (more in the coming chapters) may have had some role in initiating life on Earth.

However, the quantity of space-derived phospholipids is inconsequential compared to the massive amounts of phospholipid molecules in today's biosphere.[2] Under the influence of wave action, shaken membrane bilayers formed *protocells*, closed membrane-bound vesicles that are the rudimentary precursors of biological cells.

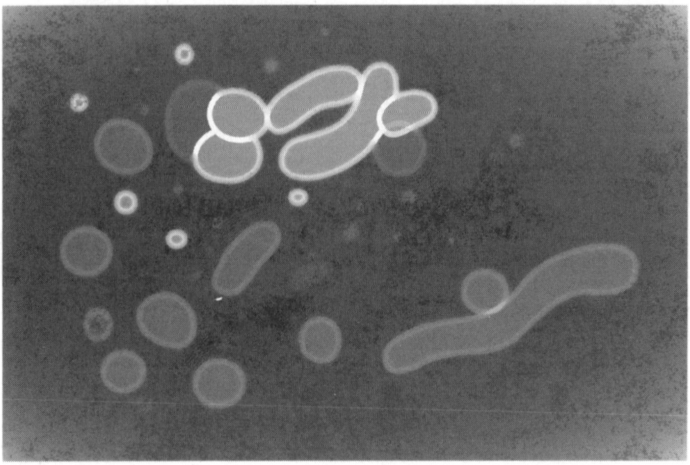

A solution of phospholipids in water, when shaken, spontaneously form bacterial-sized "protocells."

In the electron microscope image shown on the next page, the phospholipid bilayer appears as a trilaminar (three-layered) structure. This appearance is due to the water-loving hydrophilic phosphate portions of the phospholipids lining both inner and outer surfaces of the membrane that stain black; and the membrane's central hydrophobic lipid zone, remaining unstained, appears white.

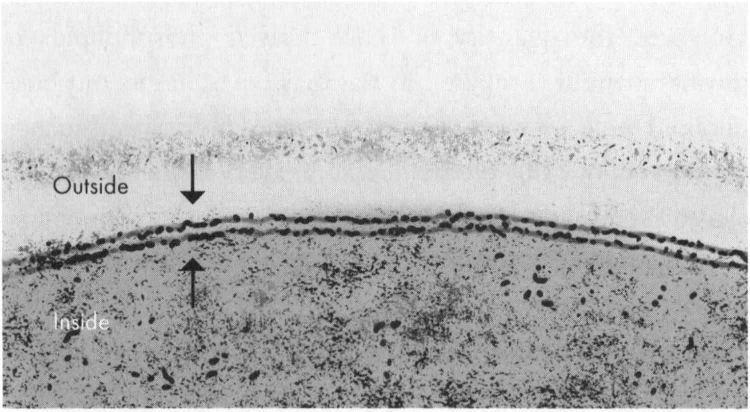

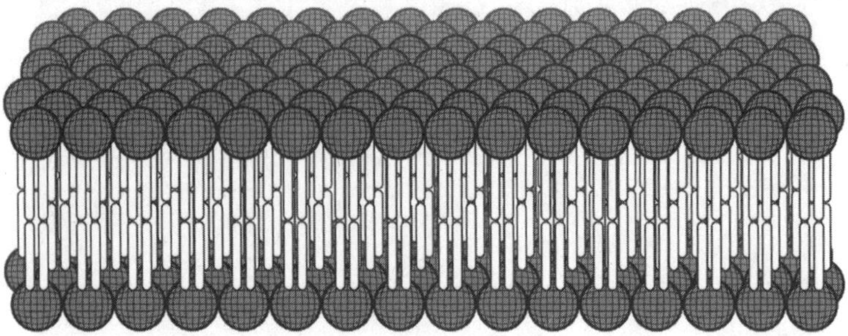

Bilayer Membrane

The figure at the top reveals the electron microscopic image of a typical cell membrane's trilaminar structure as a dark-light-dark layer forming the boundary at a cell's surface. The illustration on the bottom shows that the membrane's trilaminar appearance is derived from a bilayer of phospholipid molecules, wherein the phosphate groups of the molecule stain dark while the central opposing lipid "legs" of the molecule form the unstained (white) portion of the membrane.

During membrane formation, micro *chromophores*, crystalline molecules that absorb visible and near-ultraviolet light, become integrated into the membrane's interior hydrophobic zone. When light photons hit the chromophores, the absorbed photonic energy generates a voltage potential by separating positive and negative charges across the membrane barrier. The conversion of light energy into an electrical potential energy that spans the membrane turned the cell into a charged "battery."[3]

The membrane's central hydrophobic lipid zone serves as a barrier that protects the cell's interior by prohibiting the transport of environmental ions and molecules into the cell. How can a cell survive if nutrients cannot be absorbed and cytoplasmic waste products cannot be eliminated through an impermeable membrane barrier?

Enter a special group of IMPs, *channel IMPs*.

Like phospholipids, IMPs contain hydrophilic and hydrophobic regions. IMPs seek stability by integrating their hydrophobic regions into the membrane's central hydrophobic lipid core.

Membrane receptor IMPs possess attached polysaccharide ("sugar") chains that extend from the membrane's inner or outer surfaces and serve as antennas tuned to read environmental signals.

When an environmental signal, from *either a molecule or a nonphysical vibrational field*, resonates with the

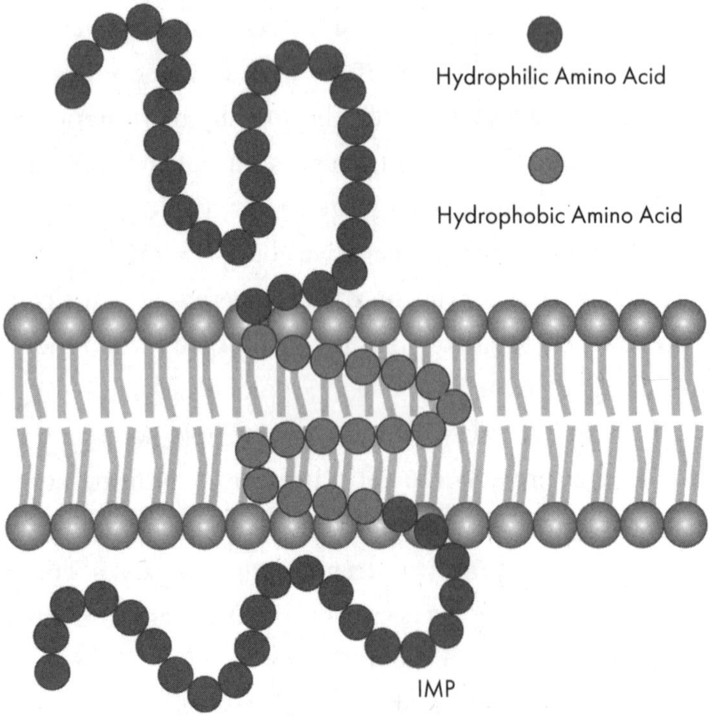

An IMP (beaded amino acid chain) is embedded in the phospholipid bilayer. The protein's hydrophobic amino acids are integrated within the membrane's central hydrophobic lipid zone. The protein's hydrophilic amino acids extend into the environment from the outer surface and into the cytoplasm from the membrane's inner surface.

receptor's polysaccharide antenna, the entanglement alters the formerly balanced charges in the receptor protein's backbone. Alteration of the charge distribution in the peptide backbone causes the receptor protein to switch from a resting conformation into an alternate "active" conformation.

Activated receptor proteins serve as signals that bind to complementary effector IMPs that in turn causes the effector protein to shift to its active conformation. Activated effector proteins have three different functions: They can become enzymes shaping metabolism, they can regulate the cell's protein cytoskeleton causing the cell to change shape and move, or they can serve as channel proteins that regulate the flow of specific ions and molecules across the cell membrane's barrier turning the membrane from a nonconductor into a *semiconductor.*

As the name implies, channel proteins form canals that allow electrically charged ions and small molecules to be transported across the membrane barrier. In its inactive "resting" state, a channel IMP's conformation is in its closed state, making the channel a nonconductor. When activated by binding with an activated specific receptor, the conformation of channel IMPs changes to create a central canal through which ions and chemical building blocks can be transported into or out of the cell's cytoplasm.

Channel proteins are highly specific for the type of ions or molecules they transport. For example, there are channels that allow the transport of ions, such as potassium, sodium, and chlorine. Other channel proteins facilitate the diffusion of small molecules of sugars, amino acids, nucleosides, and water to cross the plasma membrane. With the insertion of channel IMPs, the membrane barrier is no longer a nonconductor, nor does it conduct everything; instead, the cell membrane becomes a semiconductor.

One of the most important channel IMPs is the specialized *sodium-potassium pump* protein. When activated, these IMPs behave like a swinging door, transporting three positive sodium ions (Na+) out of the cell while bringing two positive potassium ions (K+) into the cell. The pump's activity creates and maintains a higher concentration of positive ions outside the cell and a higher concentration of negative ions within the cell. The imbalance of positive and negative charges across the membrane generates a *membrane voltage potential*, an electric charge that averages about -70 millivolts under the membrane.

When activated-channel IMPs transport charged ions, the transmembrane flow of formerly separated electrical charges across the membrane barrier depolarizes the membrane potential and creates a vibrational electric field that travels through the cytoplasm. The structure of the canals in different channel proteins causes the transmembrane ion flow to generate unique vibrational frequency signals that activate cytoplasmic proteins responsible for metabolic functions and cell behaviors.

The phospholipid molecules that comprise the membrane's bilayer have a defined thickness of approximately

8 nm (nanometers), which is eight-billionths of a meter (0.000,000,008 meter). IMP proteins measure about 9 nm in diameter, which means they cannot be stacked on top of one another in the membrane, so they are limited to forming a monolayer within the membrane. The fact that IMPs are restricted to a monolayer means that the number of IMP proteins in each membrane is directly correlated with its surface area. While a cell membrane is a three-dimensional object with a defined constant and uniform thickness, it can be modeled as a two-dimensional flat surface because the phospholipids are restricted to a monolayer.

Again, fractal modeling is useful because the most efficient way to pack a two-dimensional surface area into a three-dimensional space is through the use of fractal geometry modeling. But even with the best fractal packing, the number of IMPs is limited by the surface area of the membrane, which leads me to a way of tracking conscious evolution: Conscious evolution can be mathematically modeled by calculating the expansion of membrane surface area to accommodate as many perception units (IMPs) as possible.

Fractal modeling also provided me with the single most "aha" moments I've experienced as a scientist. One of the most important examples of fractal self-similarities is the relationship between the structure and function of the cell membrane with that of a silicon-based computer chip. To review:

Evolution

Expansion of Awareness

Unit of Perception

Evolution

Expansion of Awareness

Expansion of Membrane Surface Area

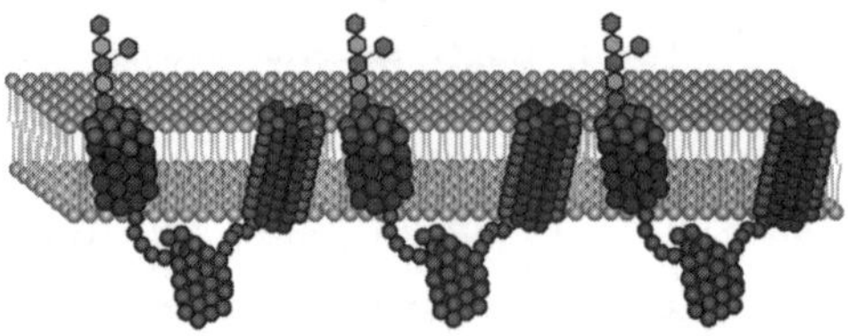

Unit of Perception

The illustration on the top shows a cross of a cell membrane. The phospholipid bilayer is approximately 8 nanometers (8 billionths of a meter) in diameter, while IMP molecules are about 10 nanometers in diameter. Consequently, additional IMPs **cannot** be stacked upon one another. The illustration on the bottom reveals that to increase the IMP population, these molecules can only be distributed as a monolayer within the membrane bilayer.

1. The phospholipid molecules that form the membrane's bilayer are organized into a *crystalline* array (the perfectly aligned "at-attention soldiers").
2. The lipid core in the membrane bilayer creates a hydrophobic barrier that prevents transport across the membrane. Normally, that would make the membrane a *nonconductor.* However, activated-channel IMPs allow selected ions and molecules to cross the barrier, making the membrane a *semiconductor.*
3. Embedded in the membrane are two functional classes of IMPs: signal receptors (inputs) and function-generating channels (outputs). In defining the membrane, a synonymous term for the receptor IMP is the word *gate.* Consequently, the embedded receptor-effector IMPs can be defined as *gates and channels.*

The membrane's molecular structure and function defined:

The membrane is a crystal semiconductor with gates and channels.

This is exactly the same definition used to describe a silicon computer chip. It is not a coincidence that the cell membrane and a computer chip share the same definition, because as "information processors," they share self-similar fractal structures and functions. This conclusion is important because it reveals that the cell membrane is NOT *analogous* (i.e., like) to a silicon computer chip; the membrane IS *homologous* (i.e., the same as) to a silicon chip!

As an information processor, the membrane converts environmental signals into specific biological behaviors: In this capacity, the membrane functions as a *transducer*, a mechanism that translates a variety of environmental signals, including physical and chemical stimuli, into electrical signals.

Receptor IMPs provide for signal "input" (I) and effector IMPs, including ion channels, generate signal "outputs" (O). Computer technology defines a membrane receptor-effector complex as a gate and channel, which in turn is referred to as an I/O, a data "bit," short for "binary digit," which represents the smallest unit of data a computer or a cell membrane can process. In a computer, processing power directly correlates with the number of bits the device can handle. Similarly, the processing power of a membrane is directly correlated with the number of receptor-effector IMP complexes representing biological data "bits."

Understanding a cell by viewing it as a computer chip led to the insight that cells are *programmable* and that the programmer lies *outside* the computer/cell. This is the same Lamarckian story revealed by the fate of cloned stem cells in my laboratory research: *Biological behavior and gene activity are linked to information from the environment, which is downloaded into the cell, a "programmable chip" whose behavior and genetic activity are primarily controlled by environmental signals, not genes.*

Chapter 8

THE QUANTUM LANGUAGE OF CELLS

A historic insight into the nature of the "language" of the cell began with the philosophy of René Descartes and Isaac Newton in the late 1600s. Collectively, their work subdivided the universe into two distinct realms: matter and energy. It was understood that matter could only affect matter, and energy could only affect energy *and* that matter and energy did not influence one another.

This assumption had a significant impact on the science of biology because, as a matter-based "machine," the body was thought to be unaffected by energy fields that included consciousness and spirituality. It has only been in very recent times that conventional medicine has considered the influence of energy, including the mind and spirit, in shaping an individual's health and behavior.

Today, quantum physics is recognized as the most valid science because its foundational theories have been corroborated by current science. Quantum physics upended the understanding of matter and energy when the science was formalized at the fifth Solvay Conference in Physics in 1927. In contrast to a dual-realm Universe comprised of matter and energy, the new physics established that the entire Universe is made from only one realm: *energy*. The former realm of matter was found to represent a "strange" form of energy.

The new reality was rooted in Albert Einstein's 1905 discovery that led to one of science's most revolutionary equations: $\mathbf{E} = \mathbf{mc^2}$. This equation emphasizes that energy (**E**) and matter (**m**) are equal. Though it's not a literal translation, Einstein concluded that "reality is merely an illusion, albeit a persistent one."[1]

This theoretical notion was formalized when physicists dissected the atom's subatomic particles—its protons, neutrons, and electrons. Once thought to be fundamental physical atomic subunits, these structures were found to be comprised of even smaller "particles" derived from a menagerie of "presumed" particles that included quarks, gluons, leptons, and photons. Subsequent research revealed that these atomic building blocks are actually powerful, nonphysical spinning energy vortices, the equivalent of nanotornadoes.

Though they were formerly perceived to be the fundamental subunits of the material realm, it is now recognized that atoms are created from immaterial, invisible energy vibrations. The profound and intellectually challenging reality revealed by quantum physics is that the Universe is comprised of only one realm—*energy*. As Einstein accurately described, the material realm is simply an illusion.

The tornado-like energy vortex at the core of an atom continuously spins. In the process, it radiates invisible force fields that resemble the ripples on a pond created by a falling pebble. When the pebble hits the surface of the water, the kinetic "energy" of the falling pebble (due to the force of gravity pulling on its mass) is transmitted into the water. The resulting water ripples are shaped by the inherent wave-like character of kinetic energy forces. Energy is expressed as "waves."

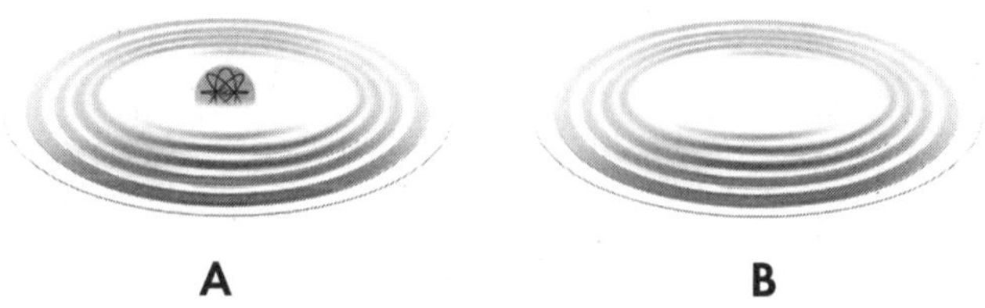

A spinning atom generates waves that resemble ripples on the surface of water. The image shown in A shows a Bohr model of an atom (with protons, neutrons, and electrons as solid subatomic particles). A more accurate model shown on the right emphasizes that the central spinning atom is an invisible, non-material energy vortex.

The "physical" parameters of the atom's ripples, their height and *frequency* (the number of wave cycles per second), are correlated with the number and distribution of electrons orbiting an atom. Each of the atoms comprising the periodic table of elements possesses a unique spectral vibrational (frequency) signature. For example, scientists can assess the molecular composition of a star by reading the spectrum of light—energy vibrations emitted by the star.

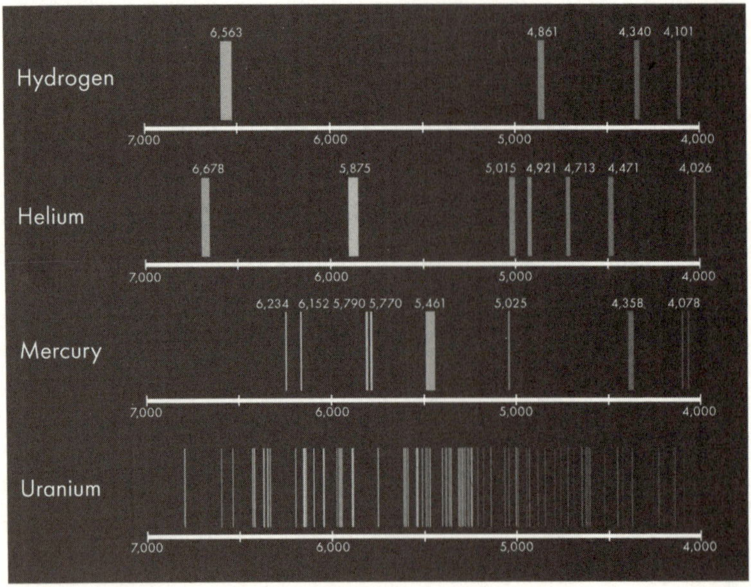

Atomic Spectrum. Each atom is comprised of a unique composition of positively charged protons and negatively charged electrons. As an atom spins, its distribution of charged subatomic "particles" generate waves of electromagnetic radiation that can be observed as specific identifying color frequencies.

Quantum physicists created energy-scanning devices that can analyze the molecular composition of specific materials and objects by assessing the energy frequencies they emit. Devices, such as CAT scans, MRI scans, and positron emission tomography (PET) scans read the energy spectra emitted by the body's tissues and organs. The energy scan illustrated below reveals the presence of a brain cancer (white) distinguished by the unique energy spectrum emitted by the chemistry of diseased tumor cells, which differ from the light gray energy emitted by surrounding healthy cells.

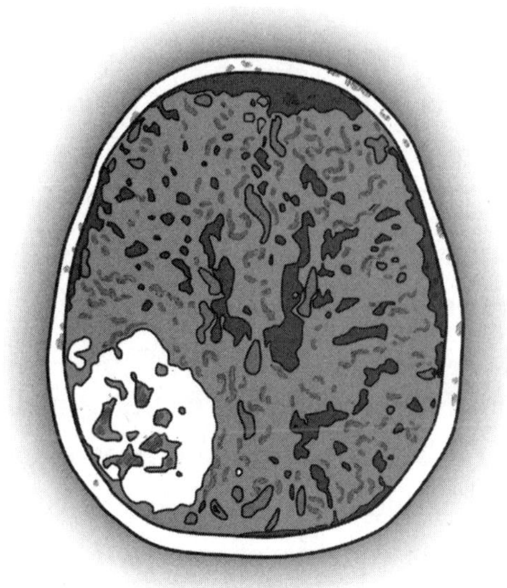

An MRI scan of a brain reveals the presence of a brain cancer stained white in the lower left quadrant. The cancerous cells are distinguished by the unique energy spectrum emitted by the chemistry of diseased tumor cells that differ from the darker energy stain emitted by surrounding healthy cells.

Returning to the pebble-and-pond analogy, consider the consequence if two pebbles are dropped into the water at the same time. The spreading energy ripples derived from each pebble converge and then interfere, or *entangle*, with one another. Where two or more ripples converge, they create composite waves that can range from energy-amplifying *constructive interference* patterns to energy-canceling *destructive interference* patterns.

Dropping two pebbles of the same size, from the same height, and at the same time, coordinates the wave action so that both sets of ripples are in *phase*, in *harmony*, wherein both sets of ripples are going up and down at the same time (A in the figure on the next page). As the ripples from each pebble converge, they overlap one another as illustrated in the schematic (B in the figure). Adding up the value of the interacting waves reveals that the height of the combined waves is greater than the height of either of the original waves (C in the figure). Wave height reflects "power," for example, a small pebble's energy makes a small, low wave, while the energy of a larger rock produces a more powerful, higher wave. The increased energy in the composite wave represents the consequence of *constructive interference*, or *harmonic resonance*.

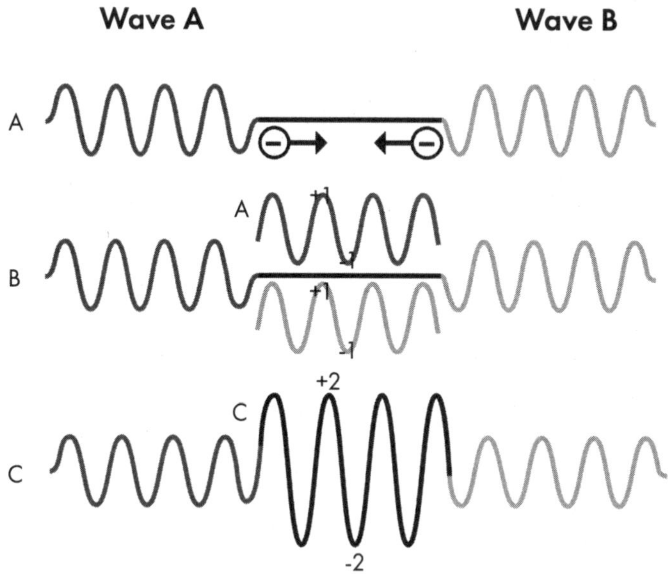

Energy is expressed as invisible oscillating waves. When two identical harmonic energy fields engage with one another (Figure A), the additive power of the resulting composite wave (Figure B) is greater than the power of the individual entangled waves. This enhancement of power is the result of a process defined as *constructive interference* (Figure C).

When dropping one pebble just before the other, the resulting waves are "out of phase," or out of sync; as the ripple of one wave is going up, the other wave's ripple is going down (A in figure on the next page). When adding up the value of the ripples at the point of convergence (B in the figure), the energy waves cancel each other. Instead of a doubling of the energy as determined by the height of the ripples, where they overlap, the water surface is calm . . . there is *no* energy wave—the energy is canceled (C in the figure). The phenomenon of canceling energy waves is called *destructive interference.*

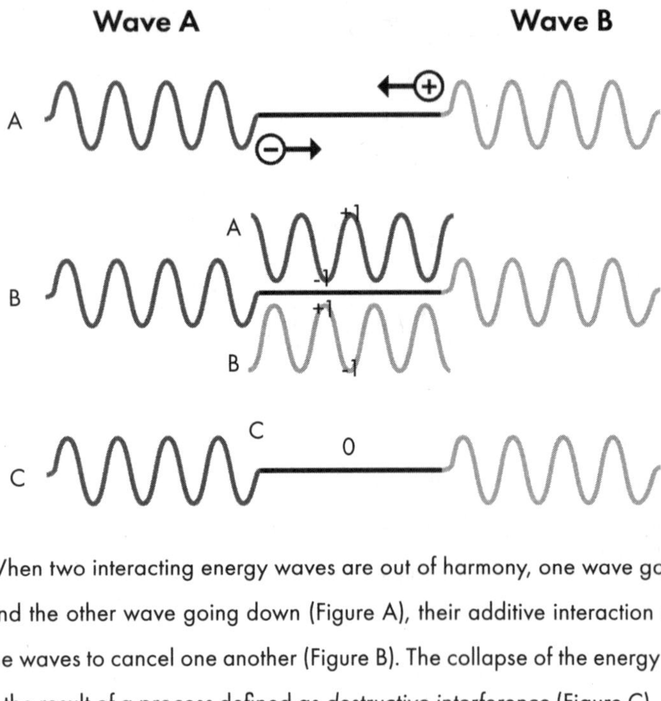

When two interacting energy waves are out of harmony, one wave going up and the other wave going down (Figure A), their additive interaction causes the waves to cancel one another (Figure B). The collapse of the energy waves is the result of a process defined as *destructive interference* (Figure C).

The spin of an atom's electrical charges generates wave patterns resembling the expanding ripples derived from pebbles dropped into a pond. As described above, the entanglement of energy waves plays an important role in biomedicine because constructive and destructive interference patterns can alter the physical and chemical properties of the cell's atoms.

The discoveries of quantum physics have also deepened our understanding of how the "magical" membrane processes environmental signals. The cell's phospholipid membrane creates a semipermeable barrier that separates a positive electrical charge on the outer surface of the membrane from a negative electrical charge on the membrane's inner surface. The resulting negative charge below the membrane is referred to as the cell's *membrane voltage potential.*

Incorporated into the membrane are glycoprotein receptors, proteins with attached chains of cyclic polysaccharides (referred to as *glycosyl* groups). The protein's polysaccharide chains extending from the membrane's outer surface are the functional equivalent of antennas that download environmental signals.

The composition of the glycosyl chain is unique for each version of glycoprotein receptors. The nature of the chemical bonds among the glycosyl molecules forming the polysaccharide chain leads to electrons coursing through the chain like race cars on a track. The speed at which the electrons cycle though the polysaccharide is expressed as a tuned *frequency*, measured as cycles per second.

When the vibration of an environmental signal is resonant, in harmony, with the vibration of a glycoprotein receptor, their energy waves entangle. Resulting constructive or destructive interference patterns induce the receptor protein to change its conformation into an "activated" form. The new receptor conformation causes the protein to couple with a complementary membrane protein channel. Receptor binding induces a conformation change in the formerly closed channel protein. In its new "active" conformation, the channel develops a central pore that connects the positively charged outer environment with the negatively charged inner environment.

Membrane channels are not all the same; they differ in their biophysical properties.

The structure of the channel determines which charged ions will pass through the membrane. There is a large family of different channel versions—for example, sodium, potassium, and chloride channels, among others. When the specific ions flow through a channel, they do not flow in a straight path through the channel's pore. Positive and negative charges built into the lining of the channel's pore cause the ions to flow through the pore in a helical pattern. When the helical-flowing charged ions exit the channel and enter the cytoplasm, they generate an electrical "spark." The electrical discharge generates a vibrational frequency whose character is determined by the nature of the transported ion and its unique helical-flow pattern.

There are multiple forms for each version of ion channels. For example, there are nine distinct types of sodium channels built into human cell membranes. While they all conduct sodium ions, they differ in the vibrational frequency their ions generate following their helical passage through the channel's pore. The character of ion-transport frequency signals is also influenced by a rhythmic opening and closing of ion channels, which produce fluctuating membrane potentials and intracellular ion concentrations.

The vibrational resonance generated by the membrane channel's flow of ions passes through the cell's cytoplasm in the same manner as "ripples" move through the water. Embedded within the cytoplasm are function-creating glycoproteins that control the cell's metabolism, motility, and behavior. When a channel's derived frequency is resonant with a function-generating glycoprotein, the constructive or destructive consequence of its signal can activate or inhibit the glycoprotein's function. This is the process by which the cell membrane deciphers environmental signals and controls the behavior and function of cells.

Energy fields have no borders, and as a result, they entangle with one another. Because of energy interference patterns, the entire Universe is one unified entangled field of energy! That makes the phrase "We are all in this together" not just a throwaway line but a concept that convinced me of the spiritual nature of the Universe. It also significantly altered the science of biomedicine because it proved that environmental energy fields, especially consciousness and spirituality, play a significant role in shaping the character of human life.

Chapter 9

ORIGINS OF THE CHEMISTRY OF LIFE

By current estimates, Earth formed 4.54 billion years ago after a rotating disc of gas and dust from a "supernova," a spectacular explosion of a star, formed the sun and planets. Driven by gravity, the dust particles and chunks of "space dust" from the supernova coalesced to form the physical planet we know today.

Early in the history of the science of chemistry, it was recognized that there are two distinct types of chemistry on planet Earth: organic and inorganic. Inorganic chemistry is associated with lifeless minerals like rocks and water. Organic chemistry is the study of the structure and properties of carbon-containing compounds that are the material foundation of living organisms. (Most organic compounds contain carbon, hydrogen, and oxygen, but they may also include other elements, for example, nitrogen, phosphorus, and sulfur.) At its origin, the molten earth was comprised mostly of inorganic minerals, which leads to the question that still baffles scientists: "From where did early life source its organic ingredients?"

In 1903, physicist Svante Arrhenius hypothesized that the first microscopic forms of life, driven by radiation from the stars, could have arrived on Earth from the seemingly bleak, hostile environment of space (the "panspermia hypothesis"). Though he had no way to test his hypothesis, research that followed concluded that Arrhenius was on the right track.

In the middle of the twentieth century, University of Chicago scientists Stanley Miller and Harold Urey hypothesized that lightning, extreme temperatures from volcanic activity, and asteroid impacts sparked the formation of organic chemistry from an atmosphere of inorganic precursors in prebiotic Earth. They decided to reproduce the conditions thought to be present in prebiotic Earth using methane (CH_4), ammonia (NH_3), and hydrogen (H_2) gases, plus water (H_2O), in an enclosed environment. Lightning was simulated by applying an electric arc. The experiment did in fact produce some amino acids and other simple compounds.[1]

Scientists later examined sealed vials preserved from the

original experiments using advanced instruments and found that more amino acids were produced in the original experiment than Miller and Urey's research reported.

More recently, researchers of astrobiology, a multidisciplinary field that has boomed since the advent of space travel, have concluded that organic compounds are ubiquitous in space, found in diffuse interstellar clouds, in dense star-forming regions, on the surfaces of minor exoplanets, and in asteroids and comets. These organic chemicals include the molecular building blocks of sugars, fats, proteins, and nucleic acids.

One example is the pristine carbonaceous astromaterial NASA's OSIRIS-REx spacecraft sampled from the asteroid Bennu and returned to Earth with in 2023. Chemical analysis revealed that among the organic carbon molecules were fourteen of the twenty amino acids used in terrestrial biology, as well as all five nucleobases found in DNA and RNA.[2]

In 1969, the same year Apollo astronauts walked on the moon, an analysis of a meteorite that landed in Murchison, Australia, found that it contained an amino acid distribution remarkably similar to Miller-Urey's discharge products. In fact, an analysis of the organic fraction of the Murchison meteorite detected over ten thousand unique compounds.

Some theorists have speculated that comets bombarding the early Earth could have provided a supply of complex organic molecules along with water and other volatiles.[3] Though the hypothesis is intriguing, given the very low concentrations of biologically relevant material, and the uncertainty about how many space-derived organic chemicals survived their fiery entry when meteors crashed through Earth's atmosphere, there would not have been enough material to account for the organic nature of the web of life's massive biosphere.

The largest percentage of Earth's organic chemistry was created later in evolution via photosynthetic processes generated by bacteria, algae, and higher plant life that contained cell organelles called *chloroplasts.* Chlorophyll, the green pigment within chloroplasts, utilizes the absorption of light to provide energy for photosynthesis, a process that employs quantum mechanical principles, in contrast to the conventional mechanics of Newtonian physics, to convert H_2 and CO_2 into organic compounds. After the hypothesized seeding of organic chemistry from space *and* the evolution of higher photosynthetic plants, planetary life was in full swing.

Astrobiologists continue to link life in space with early life on Earth by designing experiments to evaluate the question of whether organic conditions in space can support life in any form. In 2020, research conducted on the International Space Station concluded that the harsh conditions of space *can* sustain life, albeit in primitive form. Scientists at Tokyo University of Pharmacy and Life Sciences published a study in *Frontiers in Microbiology* that found that the bacterium *Deinococcus radiodurans* can survive at least three years in space even with no protection from cosmic ultraviolet, gamma, and x-ray radiation. For the experiment, a robotic arm placed a container with three panels of bacteria outside Japan's Kibo lab on the International Space Station.[4]

Meanwhile, since the 1970s, other scientists have focused on Earth's oceans as the source of life, though as University College London professor of evolutionary biochemistry Nick Lane concedes in his 2022 book, *Transformer: The Deep Chemistry of Life and Death*, "Exploration of the deep ocean has never quite matched the glamor of space." Still by his reckoning, the serendipitous discovery of vibrant life surrounding a deep-ocean "black smoker" hydrothermal vent (imagine a microvolcano) in 1977 was "on a par with the moon landings . . . This was an unimagined finding: a deep-sea food chain that apparently did not depend on sunlight or photosynthesis but lived on the gas hydrogen sulfide emanating from the vents themselves."[5]

Since then, scientists have theorized that hydrothermal vents could have provided the energy, nutrients, and a safe haven for life on Earth to begin. In 2000 in the mid-Atlantic ocean, researchers discovered a second type of vent, an "alkaline deep-sea thermal vent" whose chemistry differs from the black smoker vents studied since the seventies. Black smoker vents spew geothermally heated water (approximately 350° C) with high levels of sulfides that on contact with the cold ocean form black smoke. The chemistry of the cooler (approximately 50–90° C) alkaline vents is driven by a process called *serpentinization*, an H_2-producing geochemical reaction that has been a feature of hydrothermal systems for as long as there has been water on Earth.

When Lane assessed the chemical activity of alkaline hydrothermal vents, he concluded that quantum physics directed the step-by-step mechanics through which biochemical pathways produced organic chemistry.[6] By this theory, high heat, high water pressure, hydrogen and carbon dioxide gases, and iron and sulfide minerals, serving as catalysts, spontaneously converted H_2 and CO_2 gases into the simple organic compounds acetic acid and methane. These basic molecules provided for the synthesis of larger, more complex organic molecules including amino acids, sugars, lipids, and nucleotides, all of which are fundamental building blocks of cellular life.

Scientists, including William Martin at the at the University of Düsseldorf, who favor alkaline hydrothermal vents as the places where life likely emerged for the first time are persuaded by the symmetry between the vents' chemistry and the metabolism of early life: "The CO_2-reducing geochemistry of modern hydrothermal vents provides a model for our understanding of how such processes might have been possible at the dawn of biochemistry . . . Hydrothermal vents thus unite microbiology and geology to breathe new life into research into one of biology's most important questions—what is the origin of life?"[7]

Other scientists, including Martin J. Van Kranendonk at the University of New South Wales, have poked holes into theories about deep-sea hydrothermal vents as the origins of life and instead posit that surface-level volcanic hot springs and pools that repeatedly dry out and get wet again would be better incubators of life because: "The pools have heat to catalyze reactions, dry spells in which complex molecules called polymers can be formed from simpler units, wet spells that float these polymers around, and further drying periods that maroon them in tiny cavities where they can interact and even become concentrated in compartments of fatty acids—the prototypes of cell membranes."[8]

The question of whether life emerged from CO_2 and H_2 in deep-ocean hydrothermal vents or in surface waters where ultraviolet light energizes gases, such as cyanide, has implications for the search for life in space. If life emerged from deep-sea hydrothermal vents, the icy moons of Saturn and Jupiter (Enceladus and Europa) would be ideal places to look, because beneath their icy crust is a liquid ocean bubbling with hydrogen gas and small organic molecules judging from the plumes that jet hundreds of miles into space through cracks in the ice. If, on the other hand, light is essential for the origin of life, science-fiction-favored Mars, with its 3.65-billion-year-old hot spring deposits, would be a better place to search.[9]

Meanwhile, what Lane calls a "rift of geological dimensions" between hydrothermal vents and surface pools proponents is spurring scientists to come up with a research breakthrough, a smoking gun that will settle where (and how) life emerged on Earth.

Chapter 10

PLANETARY ETCHING AND THE FRACTAL FIELD

The following information about the role of planetary etching in the creation of life on Earth is offered as a suggestion to consider. It's a suggestion based on research and the fact that the Universe we live in is of fractal design. So, I would call the following an interesting, yet non-peer-reviewed hypothesis!

Before the appearance of life, Earth's inorganic chemistry was physically evolving for three hundred million to a billion years. While some organic molecules may have arrived from space, energy provided by deepwater hydrothermal volcanic vents and from the sun's ultraviolet and infrared light were chemically modifying Earth's inorganic chemistry.

When an energy beam hits the surface of matter, it can *etch* the physical material, corroding or eating away the surface of an object. Research on the impact an etching beam has on silicon in manufacturing computer chips buttresses this theory. One computer simulation study focused on the physical consequences that occur at the interface where the etching beam hits the silicon substrate. The computer simulation revealed a *fractal morphology* that appears at the end of the interface between a random system and a finite, etching solution. The spontaneous appearance of this morphology is directly linked to the critical slowing down of the reaction when the system is moving to a collective equilibrium.[1]

Solar photonic light beams act as an etching mechanism: The beams' photons energize a reaction in which inorganic molecules, the molecules present on pre-life Earth, interact to create the organic chemistry that is essential for life. It is reasonable to draw the inference that a similar energy field is created when a solar photonic beam etches Earth's surface and that those beams created organic chemistry, i.e., molecules that, unlike inorganic chemicals, could move and chemically interact.

Building on Einstein's quote about the all-energy, quantum Universe, "The [energy] field is the sole governing agency of the particle," it can be deduced that over time, mobile organic molecules adapt to and then manifest a physical complement to the etching's generated fractal field. The fractal design of nature then becomes a direct consequence of the organic transformation of the etching of inorganic Earth, followed by the subsequent assembly of the organic molecules into the newly created fractal field.

That leads to the conclusion that evolution is directly influenced by a two-step process induced by solar etching beams:

1. The manifestation of a fractal energy field (*creationism*);
2. The creation of organic chemistry that over time (*evolution*) assembles into the physical web of life.

Understanding this two-step process could settle (but likely won't anytime soon, given how entrenched the two camps are) the ongoing controversy between creationists and evolutionists. Creationists do not accept that there was a linear evolution of life-forms and believe instead that out of thin air (by God's decree) animals just appeared. And evolutionists do not recognize the preexistence of a patterned energy field as a precursor to evolving organisms. The fractal-etching concept unites creation AND evolution as cooperative processes in the origin of planetary life.

Chapter 11

THE NEW SCIENCE

Adding to the debate about *where* life first emerged on Earth is the debate about *when* life first emerged on Earth. The oldest known life-forms on Earth are *stromatolites*, fossilized films containing communities of diverse bacteria and photosynthetic cyanobacteria that are 3.5 billion years old.[1]

In 2024, in what the journal *Nature Ecology & Evolution* described as a "gobsmacking study," University of Bristol researchers posited that life on Earth may have emerged earlier than thought, around 4.2 billion years ago, only a few hundred million years after Earth emerged 4.5 billion years ago. The Bristol researchers came to that conclusion through a study of the genomes of organisms alive today, dubbed LUCA for the "last universal common ancestor."[2]

I know that as I write this book, researchers around the world are laboring on other studies that will deepen our understanding of evolution, some of which will be "gobsmacking," and that these studies will lead to an ever-evolving theory of evolution!

I felt the need to offer this theory of conscious evolution because my work over the last fifty years has convinced me that the evolution of civilization requires not just a science-based theory but also science-based activism.

When I say science based, I do not mean the tunnel-vision science inspired by the remarkable discovery of the double helix that has led to an unfortunate emphasis on genes as destiny. As I've argued in this book, the emphasis on genes coupled with the deeply embedded notion that evolution is a story of survival of the fittest fueled by random mutations has led to the toxic civilization in which we currently find ourselves: a morality-free and deadly competition of the fittest individuals.

Instead, what I mean by science-based activism is activism based on the revisionist research that underpins the New Science, research that demolishes the four dysfunctional "myths" I recap below.

Myth #1 *Humans are slaves to our inherited genes, making humans victims of their own biology.*

New Science Reality: Genes are regulated by signals from the environment mediated by our nervous system, which gives us the ability to create our own destinies and also makes evolution not a story of increasing numbers of genes but a story of increasing consciousness.

Myth #2 *To survive, humans need to engage in vicious, survival-of-the-fittest competitive battles—meaning that war is inevitable.*

New Science Reality: Cooperation not competition is the driving force of evolution, which means we have the ability to create peaceful societies in which war becomes obsolete.

Myth #3 *Humans are on Earth thanks to a random throw of the genetic "dice," which implies there is no reason for human existence.*

New Science Reality: *Adaptive* epigenetic changes initiate evolution, which means our lives do have meaning. We are here as "gardeners" who complement and support the health and vitality of nature.

Myth #4: *Humans live in a material world on a mission that is humorously defined as "Those who die with the most toys win." This cultural quest is responsible for humans undermining the environment in order to support their need for big cars, big houses, and a full toy box that proves they are winners.*

New Science Reality: Thanks to quantum physics, we now know that the material world is an illusion; human beings and the Universe we live in are connected and entangled in an energy field. Consequently, we can use our conscious minds to gather together in community and shift energy with love to propel the next evolutionary step, the creation of a new integrated organism, Humanity.

As I wrote *Beyond Darwin*, I looked up biographical information about Harold J. Morowitz, whose book, as I wrote in chapter 1, I stumbled on in the Stanford library thirty years ago. I learned that he died in 2016 after a long and distinguished research and teaching career at Yale University and George Mason University.

In the early eighties, Morowitz testified in the Arkansas court case *McLean v. Arkansas*, a replay of the "Snopes" trial (1925) that challenged a state law mandating the teaching of "creation science" in schools alongside evolutionary biology. Morowitz's testimony that there is no scientific basic for the creationist belief in the origin of life and therefore it should not be taught as science in public school curricula was so persuasive that the judge ruled in his favor.

I also learned that when Morowitz delivered a commencement address, he told students, "Conformity is not necessarily a virtue." (And not in nature either. A biofilm would never have led to evolution if all its bacteria had the same function; it was bacteria's diversity that enabled the evolution of cells and humans.)

Bravo, Morowitz! Here's to nonconformist scientists throughout history whose research has led to unintended consequences. That includes Darwin (despite the inadequacy of his theory) and Lamarck, because during their lifetimes *any* theory of evolution was considered heretical and challenged religious orthodoxy, as did Morowitz in Tennessee. And here's to pioneers who use the scientific method to challenge scientific orthodoxy as I did (before epigenetics had a name), when my experiments with cloned cells led me to question the dogma of genetic determinism and the mechanisms of evolution as proffered by Darwin.

Morowitz not only urged students to value nonconformists, but he also told them that "hopefulness is a moral imperative," a moral imperative I subscribe to as well, though I am not naïve. In the history of evolution, civilizations have come and gone, including the Roman Empire, where vast income inequality and a small elite with private bodyguards that protected them from the masses presaged its collapse. Sound familiar?

There is no doubt that we are at a crossroads, but rather than focusing on the eerie and accurate parallels between Western culture and fallen empires, I stress the counterintuitive portion of my theory that holds that the chaos and destruction we are experiencing can be a *good* sign. Chaos ignites change, and it can be the precursor to creating a more advanced version of civilization. It is my hope that this book provides not just an alternative theory of evolution, but a blueprint for how we can ignite that change so we can evolve (and not a moment too soon!) from a survival of the fittest to a survival of the most conscious culture that will heal ourselves and our planet.

I'll end this chapter where my science journey began: with cells. In this case, the poetically named "imaginal" cells of a caterpillar. After a period of ravenous consumption, a caterpillar finds an appropriate perch and forms a chrysalis (cocoon). Inside the chrysalis the cellular structure of the caterpillar breaks down and dissolves into organic cell soup.

Then miracles start happening. Imaginal cells that were dormant in the ravenous caterpillar begin the process of creating a new form and structure. At first these imaginal cells are viewed as threats by the caterpillar's immune system. But they persist, multiply, and ultimately connect with each other. They form cell clusters and clumps that begin resonating at the same vibrational frequency and start sharing information back and forth until they hit a tipping point. They begin acting not as discrete individual cells but as a smarter multicellular organism. The end result is that, from the ravenous caterpillar and the cellular chaos in the cocoon, a butterfly emerges as a new and beautiful non-ravenous creature with the lightest touch on the planet.

Civilization's current chaotic state resembles the chaos in the cocoon. I know that not all butterflies make it out. However, as the most evolved species on the planet, each of us can serve as an imaginal cell and band together to effect change. If we let go of the myths that have brought us to the brink of extinction, we can collectively create a more evolved and thriveable (if I can be allowed to coin a word) civilization, a social order that leaves behind the ravenous pillaging of Earth's resources and aspires to reach the next level of conscious evolution, the creation of Humanity—a beautiful, highly evolved superorganism, a caretaker that lives lightly on this one precious Earth.

Chapter 12

THE SCIENCE OF SPIRITUALITY: THE KEY TO GLOBAL HARMONY

No two human beings are biologically identical. Each human possesses a unique personal identity, the equivalent of a PIN (personal identification number).

Scientists concluded that every human being is unique when they studied rejection reactions following blood transfusion experiments and tissue and organ rejections after transplants. These studies revealed that the immune system of recipients of grafted tissues treated the transplant tissue as a foreign invader, an antigen, and destroyed it. Today's miracles of successful organ transplantation are only made possible by pharmaceutical drugs that inhibit the destructive function of the recipient's immune system.

How does the immune system recognize the personal identity that distinguishes one person from another? Immune recognition is based on identifying the molecular characteristics of each foreign antigen such as toxins, bacteria, viruses, and parasites and responding to them. Consequently, the physical markers that distinguish "self" must be present on or within our cells.

Research on rejection reactions following blood transfusions uncovered the existence of blood cell molecules called *human leukocyte antigens* (HLA) that determine whether a transfusion will be accepted or rejected. Subsequently, HLA "antigens" were found to be a constituent of *major histocompatibility complexes* (MHC I), a family of protein receptors found on the surfaces of almost all human cells, with the major exception of red blood cells. (Because red blood cells do not display "self-identity" MHC receptors, successful blood transfusions between people are possible based only on their blood type: A, B, AB, or O.)

"Self identity" is determined by specific MHC I receptor proteins on the surface of our cells' membranes. MHC receptor proteins are encoded in three genes: HLA-A, HLA-B, and HLA-C. Each of these three HLA genes codes for a single MHC receptor, and all three genes are expressed simultaneously to define a person's identity. The family of MHC I genes is also expressed *co-dominantly*, which means that the three HLA genes from the paternal chromosome and the three from the maternal chromosome are all expressed at the same time. Consequently, an individual's self-identity is determined by a composite of *six* different MHC receptors.

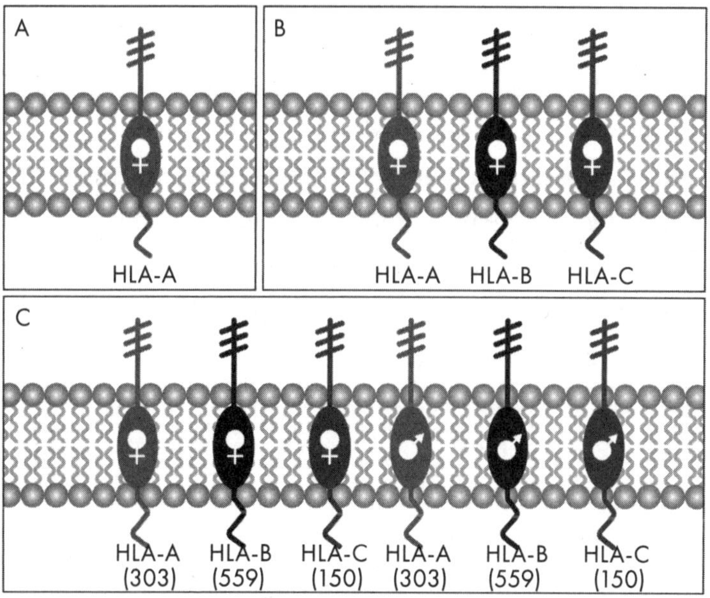

HLA membrane receptors. In panel **A**, HLA receptors are IMPs (integral membrane proteins) that extend from the membrane's outer surface. In panel **B**, HLA receptors are expressed *polygenetically*, which means that all HLA genes (A, B, and C versions) are expressed at the same time. In panel **C**, HLA genes are also co-dominant, which means that the HLA genes from both the maternal and paternal chromosomes are active at the same time.

Thus, there are six HLA genes present on almost all of the body's cells that have a nucleus. Each of the HLA genes possess a variety of alternative forms called *alleles*.

How many unique MHC I personal identity codes can be created? There are 303 different versions (alleles) of the HLA-A gene, 559 versions of the HLA-B, and 150 versions of the HLA-C gene. The number of possible six-gene MHC I codes that can be assembled from 1,012 different versions of HLA genes can provide a unique identity for over 645 *trillion* different human beings. Excluding a random accident, no two human beings share an MHC I code!

MHC I proteins are surface receptors, "antennas" that download specific environmental signals, which means that our "source," that is, our personal identity, is not contained within the body but is an awareness received from the environment by MHC receptor proteins.

What type of environmental signals are downloaded by the MHC I receptors?

In a quantum mechanical Universe, all signals are energy, whether they are physical markers or invisible waves, a fact Einstein famously summed up as: "The field is the sole governing agency of the particle."

In physics, immaterial energy fields are defined as: "invisible moving forces that influence the physical world." Interestingly, the definition of *spirit* is exactly the same "invisible moving forces that influence the physical body." Quantum physics explains how an energy field, downloaded by MHC

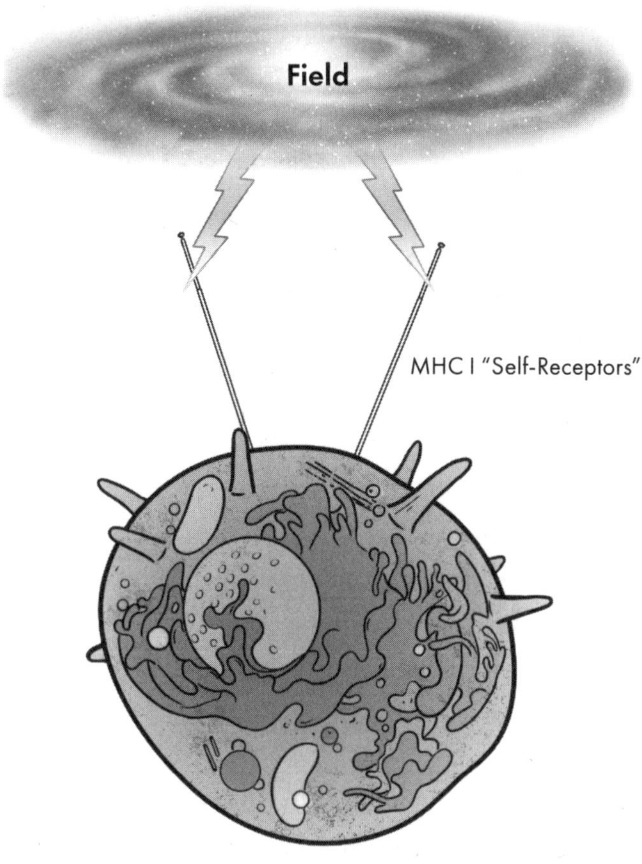

Protein receptor IMPs extending from the outer surface of a cell's membrane serve as "antennas" that respond to environmental signals. Each human possesses a unique set of MHC I receptors, defined as self-receptors, that downloads a specific identifying frequency from the environment, referred to as *spirit*.

I receptors identified as spirit, governs the character and behavior of the human body.

The source of *all* humans is derived from the same spiritual energy field. The existence of that spiritual energy field was illustrated in a science experiment involving the cells of astronaut and scientist Dr. Brian O'Leary. Some of O'Leary's cells were placed in a test tube that was moved 350 miles away from his location. A probe was introduced into the test tube to record the electrical activity of O'Leary's cells. On half of a split-screen TV was O'Leary's live image. On the other half of the screen was a readout of the probe in the test tube with O'Leary's cells. When O'Leary exhibited an emotional response, *instantaneously*, the cells 350 miles away activated an electrical response.[1] This study revealed that a nonlocal response connects the actions of a body with an immaterial energy field that many have come to know as a Spirit.

Consider the body to be a television. Our personal "show" is downloaded from an environmental "broadcast." When a TV breaks, we refer to it as being dead, but does that mean the broadcast died as well? Get another TV, turn it to the station, and the "show" is still there. This insight reveals that "we" cannot die, because "we" are derived from an environmental field, a broadcast. Our identity is actually an immortal energy field we call a "Spirit." When we die, our spiritual energy field becomes disembodied. However, when

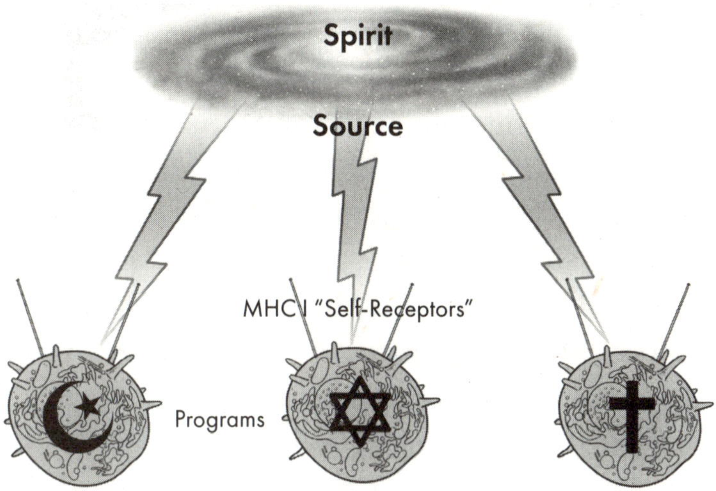

All humans download their MHC I *spirit* field from the same environmental source. However, religious programs acquired in the first seven years of life create an artificial separation among the population. This separation, often expressed in the form of a hostile opposition, has over the years frequently precipitated an endless series of wars.

a newborn possesses the same MHC I receptors we have, our Spirit is reincarnated—our same individual Spirit whether our new body is male or female, white, black, brown, red, or yellow.

All humans have a personal unique "frequency" derived from the same energy field. Unfortunately, the spiritual energy field that unifies all human souls was fractured when religious programming, including Christian, Jewish, and Muslim beliefs, subdivided the population into conflicting factions with dire consequences; until recent times, most of the wars and violence in the history of civilization were fought over opposing religious beliefs.

We may rightfully consider the collective human energy field as representing God, *All That Is*. In this case, each human is a piece of God, and we are all "creators" manifesting our personal life experiences. The sum of our collective experiences creates the character and behavior of civilization. Our evolutionary imperative is to realize that the sanctity of all humans is from one source, and that collectively, we are the equivalent of cells in a single organism. The beautiful story of conscious evolution is that once we forgo competition and collectively thrive together, we will realize the truth that Heaven is here on Earth.

PART 2

INTRODUCTION

"Knowledge is power" is an accepted truth. Another way of saying the same thing, perhaps more appropriate in today's world, is: "A lack of knowledge is a lack of power." Who, or whatever organization, controls knowledge controls power. Before Darwin, in our previous version of civilization, referred to as Monotheism, knowledge was controlled by the Church's black-frocked priests. The Jesuits enhanced the Church's power by following this adage: "Give me a child until seven and I will show you the man." As a consequence, Church schools were created to teach religious "truths," rules and laws, and download their fear-of-God, disempowering beliefs into the subconscious minds of developing children.

As I emphasized in *The Biology of Belief*, the character of our lives is controlled 95 percent by the beliefs downloaded into the programmed subconscious mind, while only 5 percent of our lives is controlled by the conscious mind, the source of our spirit, our wishes and desires.[1] In today's civilization, referred to as *Scientific Materialism,* white-coated "priest" scientists began to program "truths" into the minds of the public at the turn of the twentieth century.

Science's beliefs about our origins and the nature of life, though differing greatly from religious teachings, were equally disempowering because they emphasized the role of genes in controlling the fate of our lives and dismissed the connection between our bodies and consciousness and spirit. As a child, I bought into these beliefs that later guided me into a career as a research scientist and medical school professor. As a teacher, I passed these "truths" of scientific knowledge on to medical students, and in so doing helped shape their understanding of the practice of medicine.

My epigenetic research on cloned stem cells, cited earlier in this book, in conjunction with studies on the nature of quantum physics, profoundly altered my conscious awareness of the mechanics of life. The new awareness revealed that we are not victims but powerful creators in shaping the character of our lives. As a result of my new scientific insights, I was no longer able to teach the required curriculum emphasizing genetic determinism and Darwinian theory; I resigned my professorship and went jobless into the world. With little or no income, my personal life was in a downward spiral. Let's just say that in the early days of my self-enforced "retirement," I had so little money that I could have applied for food stamps, a big fall for a former tenured university professor.

Despite my personal situation, I was driven to bring this new self-empowering knowledge to the public. However, in my early presentations, the audience's collective response was something like, "If beliefs are so powerful, Dr. Lipton, why does your life look the way it does?" While not using these words exactly, in return I offered, "Do as I say, not as I do."

It was then, in a moment of personal embarrassment, that I realized a significant insight: Knowledge in the mind but not used in life is powerless. That is when I truly woke up and went on a mission to live my "science." The radical change in my life's circumstances provided me with the best teaching opportunity. I turned my personal and research experiences into a book, *The Biology of Belief*, whose message has now gone global in over fifty languages.

For years, people have come up to me after lectures or emailed me to tell me that *The Biology of Belief* helped them cure their cancer and save their lives. I would always respond the same way: "You have had this powerful personal experience, and YOU can be a valuable teacher to help others."

Funny where life leads. Suddenly, at the age of eighty, I received a cancer diagnosis, so now I am the one with personal experience that I can translate into biomedical science and pass that wisdom on to the public. And as it turned out, after all my years of study and implementing the lessons I learned from the New Science, I still have quite a few personal lessons to learn!

Funny also that my body's tumor is anchored in my tongue, which I have used for decades to communicate my understanding of the New Science all over the world. So why a tongue cancer? I believe it was a wake-up call. I have understood the science behind the theory of conscious evolution presented in Part 1 for the last twenty years. While the words in this book have been in the back of my mind for over two decades, I hesitated to put them into print for several reasons, among them the question, "Who am I to challenge the great Charles Darwin?" As my public lecture schedule started to build up, I used my time to prepare and present PowerPoint programs and write lecture abstracts as an excuse to put off writing the book!

Despite my long-standing hesitancy to complete this manuscript, I am also aware that the new theory I offer, predicated on recent scientific insights, significantly rewrites the mechanisms underlying the practice of conventional biomedicine and profoundly challenges the validity of Darwinian theory. The lesson I gleaned from my "belief" diagnosis is that my tongue issue is telling me to talk less and write more, which is why I'm finally finishing *Beyond Darwin*, the book that has long been roiling in the back of my mind.

As you know from the last chapter in Part 1, I do not believe that we "die" when our physical bodies stop functioning, or as Shakespeare so much more poetically put it, when we "shuffle off this mortal coil." Yet despite that understanding, dealing with a cancer diagnosis still rocked my world, forcing me to face mortality head-on. We all know that one day our physical body will "die," but we live our lives mostly in denial, and in my case that denial extended to my health. Especially because I love my life and relationship with my beloved partner, Margaret, which transitioned me from my previous life of misery (as chronicled in *The Biology of Belief* and *The Honeymoon Effect*). Just as important, I truly love the work I do spreading the empowering wisdom about *Beyond Darwin*.

Sadly, one of the first lessons I learned after my cancer diagnosis is that despite my excitement about the implications the New Science should have on medicine, despite all the exciting quantum physics research, despite much-touted research about harnessing the body's awe-inspiring and highly evolved immune system to fight cancer, the only conventional medical treatments I was offered were surgery and chemotherapy combined with radiation. The surgery would have removed my tongue, for me, a completely unacceptable surgery that would take away the instrument by which I have communicated my passion for science since I was a child. The problem with chemotherapy and radiation therapy is that it not only kills cancer cells, it wipes out stem cells needed to repair the body and interferes with the immune system's sophisticated arsenal of defenses.

In 2024, I went to Costa Rica to study with a specialist

in nutrition and engage in a new, healthier diet. My nutritionist offered this very pertinent wisdom: "The body is like a donkey, and the brain is the master. When the donkey wants to work it does so, but when it doesn't want to work it won't do it even if the master whips it to get the donkey to move."

I turned my consciousness away from the lifelong activities of my brain and started to consider the welfare of my physical body. I began to look outside the conventional medical box and assess my lifestyle choices and stressors that epigenetically may have contributed to the malfunction of my up-until-recently robust immune system. I did not consider that the problem was related to the genome I inherited, since my research revealed that genes do not control biology; gene activity is controlled by environmental signals.

It is a shock to many that there is not one gene that causes cancer, but the notion that inherited genes are responsible for most cancers, an axiom of conventional medical science, continues to fall apart under scrutiny. One British statistician told *The Telegraph*'s science editor that 70–90 percent of cancers are caused by external risk factors. Kate Allen confirmed: "Our research has shown that many cancers are caused by external factors, and that there are changes that we can all make to our lifestyles to significantly reduce our risk of cancer."[2]

As pioneering "biological" dentist Dominik Nischwitz writes in his book *It's All in Your Mouth: Biological Dentistry and the Surprising Impact of Oral Health on Whole Body*

Wellness: "Today we are fighting against different illnesses than the ones we saw a hundred years ago. These modern illnesses don't usually have just one cause for which we can take a medication. The causes are often diverse, specific to the individual, and usually only discoverable when we look at patient history and lifestyle."[3]

In the spirit of looking at history and lifestyle and as always, in conjunction with the latest science, Part 2 of this book will provide an overview of current research about lifestyle, including the food we eat, dental care, chronic stress, and their influence as the cause of disease. With all due respect to biohackers, whom I trust will teach us lots about our biology, this lifestyle review isn't about spending millions of dollars on punishing, perfectionist diets or health regimens. It's about understanding how science can point us to living healthier, lighter-on-the planet lifestyles in this quantum Universe in which matter is an illusion and we are all entangled in the web of an energy matrix.

Most importantly, in the final chapter I hope to inspire you to take to heart the evolutionary lesson of cells that banded together to create multicellular organisms, which includes the fifty trillion cells comprising the cooperative community within our bodies. If we humans can band together to fulfill the promise of Conscious Evolution, we will experience a civilization in which we humans not only survive but thrive in health, in peace, and in harmony with (not dominion over!) nature on this precious planet.

Chapter 13

BEYOND FRANKEN FOOD

Because to some extent I took my reliable immune system for granted, I have never focused on diet, which is why my books have never included diet or healthy eating advice. But in the wake of my diagnosis and studies that point to the role our industrialized food supply is playing in the rise of chronic diseases (tragically in younger and younger patients), as well as the degradation of the environment, I started thinking more deeply about the food I was raised with, the sweet tooth I have never lost, and the environmental impact of my food choices.

My parents owned a high-end supermarket that catered

to wealthy customers in Pleasantville, a suburb of New York City. Most of the customers never came to the store; they would place their orders over the phone. For this clientele, meats from our personal butchers were always "prime" grade, and we always had the best quality of fruits and vegetables (though in retrospect I can't say that they were free of pesticides!). Consequently, the food we ate at home was always of the highest quality. But despite the quality of the food I grew up with and the balanced meals I had at home, I was addicted to sugar—I ate lots of candy, cake, cookies, and other pastries. To add insult to injury, I added spoonfuls of sugar into my morning coffee.

When I went to college, I had a meal ticket to eat at the cafeteria. I only used it for a few days because the food was terrible compared to what we ate at home. I gave away the meal ticket and switched to eating quick meals at diner-like restaurants. After the first year at school, I left the dormitory, moved into a small apartment with a kitchen, and began cooking for myself—nothing too elaborate, just the basic meals I grew up with. Throughout all those life changes, I never lost my sweet tooth, so simple carbs were a big part of my diet. Meat was also a big part of every meal, including a lot of highly processed meat, i.e., bacon, hot dogs, and ham. I also loved dairy, especially milk and butter, and eggs.

As I mentioned above, shortly after I was diagnosed with cancer, I went to Costa Rica to spend six weeks with my nutritionist, which resulted in a radical change in lifestyle.

I eliminated all carbs at first and especially stopped eating sugar forever. My daily food intake included "thick" (ugh) miso every morning and evening followed by meals rich in beans and rice (each of these complex carb sources offer a unique combination of essential amino acids needed to make the body's building blocks, proteins). In fact, before the evolution of mass farming and supermarkets, beans and rice represented the major diet for people in Tibet, India, China, Japan, Indonesia, and South America. I also began supplementing my diet with a daily regimen of large doses of immune-enhancing vitamin C, vitamin D, and magnesium, among other supplements.

Though my action in radically changing my diet is new, I've long been a proponent of organic produce because of my concern about toxic pesticides used in "factory" farming. When I see photos of industrialized farmworkers spraying pesticides on edible plants while wearing hazmat suits, I shudder to think what those poisons are doing to our bodies (and minds!). Increasingly, researchers in the American corn belt are asking the same question because cancer rates among young adults are rising more rapidly there than nationally. According to a multi-authored report in *The Washington Post*: "At the turn of the century Iowa ranked 28th in the nation for cancer rates among adults under 50. Today it is 5th."

In a 2024 report, the University of Iowa's Environmental Health Sciences Research Center described Iowa as a "hotspot for environmental exposures to cancer-causing agents." While Iowa is famous for its fertile soil, the center noted that its soil contains some of the nation's highest natural radon levels, and its groundwater has the country's highest average nitrate pollution, largely due to decades of fertilizer use.[1]

I've also long been concerned about the environmental damage and the lack of nutrients that result from industrialized monoculture farming, the practice of growing a single-species crop season after season over a large area, a technology that has eliminated thousands of square miles of natural woods and fields. In 2023, Planet Wild, a Berlin-based global community whose mission includes rewilding landscapes degraded by monoculture farming, concluded: "Monocultures encourage the use of synthetic herbicides and pesticides which can kill native insects and animals—and get into our food chain."

Monoculture farming spurs the growth of harmful organisms, such as insects, rodents, weeds, and microorganisms that negatively impact the lives of humans and livestock, which in turn, leads to the use of more pesticides.[2]

In addition, humans have aggressively eliminated predator species like wolves and lions because of their impact on the farming of sheep, cows, and other forms of livestock. But this behavior misunderstands the profound role these animals play in nature. When they hunt, these animal predators serve a vital function; they only prey on the weakest individuals in a herd. This "natural selection" process is crucial for strengthening the herd by eliminating the weak individuals from breeding, thereby diluting the offspring from the strongest and best adapted individuals.[3]

Lethal humans include trophy hunters who embark on quests to kill the *strongest* animals in a herd, the ones with the biggest horns, antlers, and tusks. By eliminating the strongest, most fit organisms from the breeding community, these hunters threaten the continued existence of wild herd animals in Africa.

Another example of humans undermining the balance of nature is in overhunting whales. Whales have provided a valuable resource of oil, meat, and other products for human consumption, but thanks to human hunters, their population has dwindled dramatically. The diminishing number of whales has disrupted ocean ecosystems, which require a stable equilibrium (homeostasis) to sustain its flora, fauna, and chemistry. Instability arises when organic waste settles on the ocean floor and is no longer available to feed the fish and wildlife that primarily inhabit the ocean's surface. Whales' deep dives have significant ecological consequences, primarily through the "whale pump" phenomena when they make deep dives from the surface to the floor of the ocean and right back up to the surface. Those dives create the equivalent of a jet stream that draws bottom nutrients to the surface to support the growth and survival of the ocean's web of life. Fewer diving whales means fewer surface nutrients, which in turn, means a loss of fish and other forms of sea life.[4]

In addition to environmental devastation, evidence is piling up about a dramatic rise in chronic diseases, including obesity, cancer, type 2 diabetes, and cardiovascular disease. For example, according to the World Health Organization (WHO), obesity has more than doubled worldwide since 1990. Among adolescents, it has *quadrupled*.

WHO cites these "key" and, I have to say, depressing facts:

- In 2022, one in eight people were living with obesity.
- In 2022, an estimated 2.5 billion adults were overweight, and 890 million were living with obesity.
- In 2022, over 390 million children and adolescents aged five to nineteen years were overweight, including 160 million were living with obesity.
- Perhaps the most depressing, in 2024, 35 million children under the age of five were overweight, according to 2024 statistics.[5]

Research on the connection between the rise in industrialized food and the rise in chronic diseases is mounting. Much of the research has centered on corporate-produced, ultra-processed foods (UPFs). Before World War II, most of the food in the United States was minimally processed, but since the 1950s, UPFs have become ubiquitous. They now make up to 60 percent of the average diet in the UK and

US, with other countries on a steadily increasing trajectory as well. What are UPFs? The definition first proposed by a team of Brazilian researchers headed by Dr. Carlos Monteiro, a professor at the University of São Paulo, describes UPFs as food that lists more than five ingredients on the label, although scientists and public health advocates have been debating that definition ever since.[6]

The American Dental Association (ADA) is lobbying for a definition that explicitly includes "added sugars" because, "From a dental perspective, there is no threshold of sugar intake that is risk-free for dental decay." The ADA's proposed definition: "Industrial creations reformulated with little if any whole foods, often additives, and containing large amounts of added sugar and salt."[7] When you become a careful label reader, as I have become, you realize that the ingredients of a large number of processed foods marked as "health foods" include copious amounts of sugar to make their product taste better.

Monteiro provided an updated version of the definition of UPFs to Stanford Medicine: "They're basically industrial formulations. They contain ingredients you wouldn't find in a typical kitchen—things like emulsifiers, colorings, flavor enhancers, bulking agents, gels. And the other thing is that ultra-processed foods tend to be, in the majority of cases, higher in saturated fat, salt, and sugar—which are the three categories of ingredients we know we should eat less of."[8]

In his book *Ultra-Processed People: Why We Can't Stop Eating Food That Isn't Food,* Chris Van Tulleken, an associate professor at University College London with a PhD in molecular virology, offers a simpler, more colorful UPF definition: "If it's wrapped in plastic and has at least one ingredient that you wouldn't usually find in a standard home kitchen, it's UPF."[9]

UPF's harshest critics say they deserve to share the moniker "Franken Food" with genetically modified food because they bear little relation to the food our ancestors counted on to survive. Van Tulleken, who notes that his own children are enthusiastic consumers of UPFs, provides this devastating description in his book:

> We've started eating substances constructed from novel molecules and using processes never previously encountered in our evolutionary history, substances that can't even be called 'food.' Our calories increasingly come from modified starches, from invert sugars, hydrolyzed protein isolates, and seed oils that have been refined, bleached, deodorized, hydrogenated—and interestified. [That is, rearranging the fatty acids of a fat molecule.] And these calories have been assembled into concoctions using other molecules that our senses have never been exposed to either: synthetic emulsifiers, low-calorie sweeteners, stabilizing gums, humectants, flavor compounds, dyes, colour stabilisers, carbonating agents, firming, bulking—and anti-bulking—agents.[10]

To grapple with the issue of rising obesity rates in Brazil, Monteiro's team analyzed typical household food expenditures. To their surprise, they found that Brazilians were buying *less* salt, sugar, and cooking fats and oils. However, they also discovered that people were replacing home-cooked meals with ready-made foods, drinks, and snacks that were high in the fat, salt, and sugar they were trying to eliminate. Monteiro's team hypothesized that these foods were displacing meals prepared with fresh ingredients and contributing to the rise in obesity.[11]

A small, randomized study published in *Cell Metabolism* supported that hypothesis when it found that over a two-week period, participants assigned to an UPF diet ate about five hundred more calories per day and gained about two pounds more than those on an unprocessed diet. Researchers concluded: "Limiting consumption of ultra-processed foods may be an effective strategy for obesity prevention and treatment.[12] In addition to health concerns, the environmental costs of packaging UPF foods are high. The Coca-Cola Company, PepsiCo, and Nestlé have been in the top four global plastic polluters every year since the nonprofit Break Free From Plastic began its audit six years ago. Mars, maker of the beloved (and I can attest, addictive) candy bar Snickers is number six in the latest ranking.[13]

Environmentalists have long pointed to meat and dairy consumption as major contributors to greenhouse gases, which is one reason I'm happier with my new diet of a drastically lower intake of meat and dairy. While people are concerned that a plant-based diet is too low in protein, I found this to be a false belief when studying how well the nutritional profile of plant foods stacks up against meat. For example, in addition to more protein than meat (24 vs. 20 grams), lentils have seven times more calcium and thirty-four times more folic acid.

LENTILS VERSUS MEAT

Nutrition	Lentils (100 grams)	Meat (100 grams)
Protein	24 grams	20 grams
Cholesterol	0 milligrams	60 milligrams
Saturated Fat	0.2 grams	4.1 grams
Calcium	70 milligrams	10 milligrams
Magnesium	130 milligrams	15 milligrams
Iron	8 milligrams	2.5 milligrams
Sodium	6 milligrams	70 milligrams
Folic Acid	170 micrograms	5 micrograms

Comparing the nutritional value of plants, in this case, lentils, with that provided in meat.

In his book, *In Defense of Food: An Eater's Manifesto*, Michael Pollan famously offered this good advice: "Eat food. Not too much. Mostly plants."[14] This is great advice according to Dalia Perelman, a research dietitian at Stanford Prevention

Research Center, who notes that health professionals have been preaching it for a hundred years. Although, she concedes, that for consumers who eat UPFs, it can be next to impossible to follow because, she says, the scientists who produce UPFs are very good at their jobs: "They formulate the taste of the food to what is known as the 'bliss point.'" The bliss point is the opposite of bliss for those who struggle with overeating because it is the precise combination of sugar, salt and fat that maximizes taste and triggers over consumption. "When you hear a commercial about how you can't eat just one, they mean it," Perelman told Stanford Medicine.[15]

In an article entitled "Ultra-processed food addiction," published in the journal *Addiction*, which looked at a nationally representative sample of older adults in the USA, the authors acknowledge that middle-aged adults, especially women, are far more likely to be addicted to ultra-processed foods than older generations. They also note that the marketing of diet-focused processed foods in the 1980s may have played a major role. In the study, food addiction was linked to poor health, weight issues, and social isolation, highlighting long-term UPF risks. And finally, they warn that children today could face even higher addiction rates in the future.[16]

Van Tulleken describes the production of UPFs as a Darwinian struggle among profit-hungry corporations. You already know my attitude about describing Darwinian evolution as a competitive struggle rather than one of increasing cooperation, but as a description for industrialized food production, it is apt: "These foods have been put through an evolutionary selection process over many decades whereby the products that are purchased and eaten in the greatest quantities are the ones that outsell the rest in the market." He describes humans as the "prey" in this Darwinian production process. After production, both adults and children become further prey, victims of "predatory" advertising that includes cartoon characters designed to lure in children.[17]

Some believe that banning UPFs would be the correct policy move, though many public health advocates, including Van Tulleken, do not. Instead, they advocate "front-of-the-pack" nutrition labeling (FOPNL) warning consumers of harmful products high in calories, sugar, salt, and fat. These labels are catching on all over the world, especially in South America. All thirty-five countries in the World Health Organization's Region of the Americas (AMRO) are considering, or are in the process of adopting, the warning labels, and seven countries (Argentina, Chile, Ecuador, Mexico, Peru, Uruguay, and Venezuela) have already implemented FOPLs.[18]

Mexico offers a road map for the successful implementation of the FOPNL program, despite delay tactics by the food industry, efforts that bear a striking similarity to the tobacco industry's efforts to thwart regulation. Mexico's program is centered on black octagon labels with white text that read: "Exceso Calorias, Exceso Azucares, Exceso Grasas, and Exceso Sodio," which translates as "excessive calories, excessive sugars, excessive fats and excessive sodium."

The large size of these FOPNL labels is carefully defined, and no cartoon characters are allowed. Many companies are complying with the regulations during the first phase of its implementation by reformulating their products to decrease the number of warning logos their products must display, as well as removing cartoon characters from the packaging.

However, there are also industry efforts to exploit loopholes in the regulation. For example, some companies designed packaging labels with double "front" labels, one with the mandated warning label(s) on the real front of the package, and the other a "back" label redesigned to be a copy of the front label, but without the octagon warnings. As for the cartoon characters, they are off the product's labels but not off social media and digitalized marketing ads. In 2021, Kellogg found a way around the no-cartoon-figures-on-labels regulations by hosting a drone light show in Mexico City, where its most popular characters, including El Tigre Toño (Tony the Tiger), were displayed and livestreamed on Instagram for half an hour.[19]

While the industrial food industry is not going quietly into the night, public health advocates stay committed to efforts to lessen consumption of UPF products because research linking their overconsumption to disease is still accumulating.

One study published in *BMJ* (the *British Medical Journal*) found that men who ate high quantities of UPFs had a 29 percent higher risk of developing colorectal cancer compared to men who didn't.[20] Researchers believe that's because UPFs are generally low in the fiber the gut biome needs to thrive, and because their additives can also disrupt the gut microbiome and weaken the intestinal barrier.[21] A study published in *Gastroenterology* found that participants who ate food with a widely used emulsifier, carboxymethylcellulose, experienced alterations in their intestinal microbiota.[22]

An overview of forty-five meta-analyses reporting on nearly ten million study participants, published in *BMJ*, provided more evidence that we are paying a high price for relatively cheap and time-saving "food." The review found "convincing" evidence that a diet high in UPFs increases the risk of death from cardiovascular disease by 50 percent and the risk of anxiety disorders by 48 percent. The study also found "highly suggestive" evidence that greater consumption of these foods increases the risk of obesity by 55 percent, sleep disorders by 41 percent, type 2 diabetes by 40 percent, early death from any cause by 21 percent, and depression by 22 percent.[23]

One sobering thought for me was a University of Illinois Chicago study on the effects of sucrose and high-fructose corn syrup (HFCS) in rats that concluded: "Consumption of added sugars, particularly HFCS-55, negatively impacts hippocampal learning function, metabolic outcomes, and neuro inflammation when consumed in excess during the adolescent period of development."[24]

It's not all bad news! A new University of California at

Irvine study offers some good news, especially about my new plant-based diet! Researchers found that inulin, a dietary fiber abundant in vegetables, including onions, garlic, and artichokes, changes the gut microbiome in mice to promote better fructose metabolism, thereby mitigating or reversing "insulin resistance, hepatic steatosis (fatty liver) and fibrosis," and also helps the liver prevent inflammation.[25]

My favorite study, in part because it was conducted by a precocious sixteen-year-old and in part because it echoes my experience, was presented at the American Heart Association's Hypertension 2022 Scientific Sessions in San Diego. While dual enrolled at Broward College in Davie, Florida, and Cooper City High School, Maria Balhara gathered data on how frequently adolescents consumed twelve varieties of UPFs during eight weeks. Her study suggested that UPFs like candy, pastries, and frozen treats serve as "gateway" foods that may lead to unhealthy teen eating, which was certainly the case for me. Displaying a wisdom that I most certainly didn't have at her age, Maria concluded: "Ultra-processed foods are designed to be hyper-palatable or engineered to be as addictive as possible. They're also cheap and convenient, which makes them hard to resist. Most people are eating too many of these foods without realizing it."[26]

Amen from me!

Chapter 14

MORE THAN A SMILE: DENTISTS AS HEALERS

Since my diagnosis, I've given more thought to dentistry than I have in my entire life, which is unfortunate because if we had a medical system focused on prevention as well as treating disease, oral health would loom large, and medical and dental schools would collaborate closely rather than being firmly ensconced in their separate ivory towers.

Some "biological" dentists like Dominik Nischwitz, who received his doctor of dental surgery degrees from Eberhard-Karls University in Tübingen, are working to close the gap. Nischwitz concludes his book *It's All in Your Mouth* with a cri de coeur chapter called "The Case for a New Dentistry" that includes these "dreams": "I dream of a world where dentistry and general medicine do not form separate spheres but work hand in hand. I dream of a medicine that rejects the idea of separating the body into sections and instead acquires more knowledge about integrative concepts. I dream of a medicine that trains both patients and doctors to understand our organism as a whole."[1]

So do I, and in fact, the New Science dictates it! Separating medicine and dentistry is a relic of the old, Newtonian model of medicine, where organ systems are studied as individual entities, and teeth were more or less considered as part of the musculoskeletal system, whose physiology was not connected to the other organ systems. In that paradigm, separating medical and dental schools made sense. However, that separation is antiquated given the discoveries of New Science that support wholistic, body-mind-spirit medicine focused on the connections among all the body's organs and tissues.

Of course, Chinese medicine's practitioners were way ahead of Western medicine in this regard; for over four thousand years, they have recognized that the body's energy meridians determine the flow of information, and that there is a connection between specific teeth and specific organ systems. Translated into modern medicine-speak, acupuncture meridians, writes Nischwitz, are "points of the autonomic nervous system that are particularly amenable to stimulation." More about the autonomic nervous system in the next chapter.

Nischwitz tells the story of one twenty-four-year-old patient who came to his dentist chair as a shadow of his formerly formidable physical self. Doctor after doctor had told this patient that his "mysterious" symptom was "psychosomatic." And it was an odd symptom in that every night, at the same hour, he would wake up and have to vomit. When Nischwitz explored his patient's mouth, he found a root canal at the colon meridian with a giant cyst at its tip. After Nischwitz removed the tooth and the cyst, his patient got through his first night without vomiting. With that treatment and some lifestyle adjustments to facilitate the removal of toxins related to his job, soldering computer boards without protective equipment, he got back to competing in professional judo matches at the high level at which he had been competing before his root canal.[2]

In my case, I believe the growth in my tongue was a long-term consequence of a nerve infection caused by the extraction of an impacted tooth near the tumor site, over a year before the growth was recognized. The root canal exposed a site that then developed an abscess.

In their book, *The Roots of Disease: Connecting Dentistry and Medicine*, Robert Kulacz, DDS, and cardiologist Thomas E. Levy, MD, argue that "all root canal treated teeth are still infected or will soon become infected after the procedure is completed, even though the supposed goal of the procedure is to eliminate infection," and that those infections can lead to serious degenerative diseases, including cancer and heart disease. They also include prominent physician Dr. John Hunter's jarring description of a tooth that has undergone a root canal treatment: "A devitalized tooth (root canal treated tooth) is like a sea of sepsis under a mausoleum of gold."[3]

It's clear that iatrogenic (treatment-caused) damage is an issue in dentistry just as it is in medicine. Biological dentists are in the forefront of decreasing such damage by developing gentler, less toxic practice with safer techniques. For example, they are committed to metal-free dentistry and have established safe procedures for removing toxic mercury amalgam fillings and using ceramic for implants.

More enlightened and accessible dentistry is needed in the wake of a slew of studies connecting oral health with specific disease states. Chinese researchers looked at the connection between oral health and "chronic" disease that they argue accounts for 74 percent of global mortality, including diabetes, and rheumatoid atherosclerotic and vascular diseases. They concluded: "Chronic diseases and oral health are intricately linked . . . Addressing oral hygiene is not only crucial for dental health but also plays a significant role in reducing the burden of chronic diseases."[4]

Two recent studies in the *Neurology Open Access* journal offer more evidence that oral health is inextricably linked to systemic disease. Researchers found that people with gum disease *and* cavities were 86 percent more likely to suffer an ischemic stroke [the most common type of stroke, caused when an artery in the brain is blocked by a blood clot] than those with good oral health. Poor oral health also raised the risk of heart attacks and other cardiovascular conditions by more than a third. In a separate study, the same scientists found that adults with gum disease may be more likely to show damage to the brain's white matter, which affects memory, thinking, and balance. Dr. Souvik Sen, chair of neurology at the University of South Carolina and lead author in both studies, writes: "Taking care of your teeth and gums isn't just about your smile, it could help protect your brain."[5]

There were visionaries in previous generations who also posited the link between oral health and body health. Nischwitz opens his book with an anecdote about an unorthodox soccer coach, who somehow intuited the importance of oral health on physical health. In preparation for the 1958 World Cup, Vicente Feola, coach of the national Brazilian team, decided to send his team not just to a psychologist but also to a dentist. Of the 470 teeth examined by the team dentist, thirty-two were diseased or inflamed and had to be taken out. Though the coach's decision to send the team to the dentist was considered highly eccentric at the time, the team that no doubt had the healthiest teeth in the tournament won a decisive victory.

Canadian-born dentist Weston Price, who practiced in the United States, was another forerunner in integrating oral health with bodily diseases. At the turn of the twentieth century, Price connected the bad oral health of his patients with their increasing number of chronic diseases. Price traveled around the world to find populations as isolated as possible and came away with the insight that wherever people were eating the same way their ancestors did, their teeth were healthy and straight, and despite the fact that they had no dental health regimes, seemed immune to tooth decay.[6] Thanks to Egypt's arid climate and burial practices, twenty-first-century researchers have learned a similar lesson from ancient skeletons and mummies. Dental caries was "notably infrequent" among ancient Egyptians, and this was thought to be "due to a lack of fermentable carbohydrates in their diet impeding plaque buildup."[7]

The role the usual suspects play in poor oral health—i.e., poor diet, especially sugar intake, smoking, excessive alcohol, and poor dental hygiene—has been supported by current research. But the notion that a healthy mouth is a clean mouth low in bacteria has been upended: "In reality our mouths are only healthy when they are teeming with thriving microorganisms . . . Stability and balance among the different types of microorganism are more important than sterility," writes Nischwitz.[8]

Sound familiar? There has been a lot of well-deserved focus on the gut-brain connection and the impact of the overuse antibiotics that has led to the decreasing diversity of the intestinal microbiome. I wrote about that link in the tenth-anniversary edition of *The Biology of Belief* when I cited

Dr. Martin J. Blaser's 2014 seminal book *Missing Microbes: How the Overuse of Antibiotics Is Fueling Our Modern Plagues.*[9]

There has been much less attention on the influence of the oral microbiome, but that is changing. A 2024 study in the journal *Microoganisms* describes the oral microbiome as a "diverse and complex ecosystem that plays a significant role in maintaining oral and systemic health." Imbalances in the oral microbiome not only lead to dental problems like caries, gingivitis, periodontitis, oral candidiasis, and halitosis but also have "profound effects on systemic health, including gastrointestinal disorders, cardiovascular disease, endocrine imbalances and neurological conditions."[10]

Research is just beginning to zero in on the effect of antibiotics on the oral microbiome population. A study published in the *International Journal of Oral Science* reviewed the consequences when researchers added a "cocktail" of four antibiotics to the drinking water of mice. As expected, the antibiotic-laced water decreased diversity and dysbiosis in the gut microbiota, but the study also demonstrated for the first time that antibiotics "can disturb the oral microbiota and result in aggravated periodontitis."[11]

Alessandra Amato sounded the alarm about the overuse of antibiotics by pediatric dentists in a recent editorial in the journal *Microorganisms.* Amato found that antibiotics are "widely prescribed in dental practices for children with estimates suggesting that 66.4% of dental prescriptions in

England are antibacterial drugs." Those prescriptions were ordered for "nonindicated clinical conditions" and consequently exposed children to a "multitude of potential adverse reactions." Regarding the little-studied oral microbiome, there are warning signs about the use of antibiotics in "children as young as four years old found to harbor multidrug-resistant bacteria in their oral cavities." In conclusion, Amato makes a plea for "more concrete and definitive guidelines for managing orofacial bacterial infections and the prescribing of antibiotics in pediatric dental patients.[12]

Yikes! Though I went for annual checkups, I did not understand the ramifications of the dental work I had (several root canals) or how strongly the connection is between oral health and disease starting at a young age. Thanks to my inconveniently placed lingual tumor, I now understand this dental axiom: "Anyone looking for a solution to improve their health should start with their mouth."[13]

Chapter 15

OUR SABER-TOOTHED TIGERS

Evolution has presented us with a fantastic system to avoid being eaten by a saber-toothed tiger, the "fight or flight" response first described by Walter Cannon in 1914. The response is managed by the autonomic nervous system (ANS), which governs involuntary bodily functions and includes three major components: the sympathetic nervous system, S/ANS (protection-oriented behavior); the parasympathetic nervous system, PS/ANS (growth-oriented behavior); and the enteric nervous system, E/ANS (gastrointestinal control). "Together, these systems maintain physiological stability by constantly monitoring internal and external conditions and adjusting responses to maintain homeostasis."[1]

When the ANS recognizes the threat posed by a hungry tiger, it activates the hypothalamus-pituitary-adrenal (HPA) axis, a hormone-based communication system between our brains and the release of stress hormones by our adrenal glands. Here's how it works: The hypothalamus secretes corticotropin-releasing factor (CRF), a regulatory hormone that travels to the pituitary gland; CRF activates special pituitary cells that release adrenocorticotropic hormones (ACTH).

ACTH hormones make their way to the adrenal glands where they serve as the signal to turn on the secretion of "fight-flight" adrenal stress hormones, especially cortisol. Those hormones coordinate the function of the body's organs to provide great physiologic power to deal with threats from the proverbial saber-toothed tiger:

1. In the normal resting state of the body, blood is preferentially found in the abdomen, where its energy is used to support the organs that maintain the health and integrity of the body. In a fear state, stress hormones constrict the blood vessels in the gut, forcing the energy-providing blood to preferentially flow into the arms and legs to nourish the energy requirements so our adrenalized selves can run faster

than we ever thought possible. The stress-activated constriction of the blood vessels gives the sensation of "butterflies in the stomach" or the feeling of being queasy.

2. Stress hormones specifically shut down the immune system that protects us from threats under the skin, like harmful viruses, bacteria, and parasites. An active immune system requires lots of energy, which is why you stay in bed when you're fighting the flu. However, when facing the life-and-death threat posed by the saber-toothed tiger, the body gives up the fight against internal threats in favor of the bigger threat to survival. Stress hormones are so effective in shutting down the immune system that they are used therapeutically by transplant doctors so the immune system of a recipient of a transplanted organ will not launch an attack and destroy the new "invader" organ.
3. Stress hormones constrict blood vessels in the forebrain to redirect blood to the instinctive, faster-processing hind brain because the conscious brain is too slow to deal effectively with the immediate crisis. That's why some people labeled "heroes" when they save people from fires or car accidents reject the label by explaining that they don't remember making any decision to help; they simply acted out of hindbrain instinct.

After the threat is over and we've survived the saber-toothed tiger or its modern-day equivalent, like a near-miss car accident, once we feel safe again, we shift away from saber-toothed-tiger *protection* mode back into health-enhancing *growth* mode. When blood flow returns to the viscera, its tissues and organs return to their role of generating energy, repairing the cells, and reengaging the immune system to once again deal with internal threats. That shift back to growth mode is crucial because, as I learned in the laboratory, cells in culture dishes retreat from a hostile environment and move toward a nurturing environment. In moving between growth and protection behaviors, these modes are primarily mutually exclusive behaviors: Cells can't run backward and forward at the same time, and we can't be in *optimal* growth and *optimal* protection mode at the same time.

When the shift works well, we are paragons of stress resilience. In his review article in the *Malaysian Journal of Medical Sciences*, Dr. Mohd Razali Salleh describes the optimal shift: "Fortunately, under normal circumstances, three minutes after a threatening situation is over and the real or imagined danger is removed, the fight-or-flight response subsides, and the body relaxes and returns to its normal state. During this time heart rate, blood pressure, breathing, muscle tension, digestion, metabolism and the immune system all return to normal."[2] And I would add that the butterflies in our stomach dissipate as blood flows back to the viscera and blood flows to our prefrontal cortex so we can think rationally about our experience.

Why would nature create a "protection system" for survival challenges? The simple answer is that when humans evolved, there were very limited threats challenging survival. If you can outrun the tiger, after some minutes, the body could return to normal growth functions. However, in today's world, threats to survival are chronic, lasting 24-7-365. Simply put, humans were not designed for today's world of long-term stress.

When the system designed to deal with the acute threat of a saber-toothed tiger must deal with chronic stress, problems arise, explains Salleh: "If stress persists after the initial fight-or-flight reaction, the body's reaction enters a second stage. During this stage, the activity of the protection-controlling sympathetic nervous system declines, and adrenaline secretion is lessened, but corticosteroid secretion continues at above normal levels." Research has concluded that "almost every system in the body is adversely influenced by chronic stress. When chronic stress goes unreleased, it suppresses the body's immune system—a behavior that will ultimately manifest as illness."[3]

Duke University researchers discovered a strong link between high stress in children and adverse health conditions for them later in life. Using data from the Great Smoky Mountains Study, a longitudinal study of child psychiatric disorders that began in 1999, researchers concluded that a stable, financially secure home is essential for a healthy childhood, and that when they don't have that, their ability to handle stress carries over into adulthood. Coauthor Herman Pontzer, Duke professor of evolutionary anthropology and global health, says: "We've had an idea for a long time, since the '80s at least, that when children have adversity in their lives it affects how their bodies work, not just psychologically but also physiologically. It gets underneath the skin, and it becomes embodied in the way your body handles stress."[4]

Personally, my childhood stressors were not financial but instead stemmed from being somewhat of a social outcast. I grew up in the late forties and early fifties during the McCarthy period that blamed Jews (especially in Hollywood), and specifically Russians, as the problem creators in American democracy. I came from a Jewish family, and my father, born in the Ukraine, still had a strong Russian accent. Neighbors would not let me play with their children, fearing my "Bolshevik" upbringing would be a bad influence. In school, despite my love of learning, especially science, a major stressor was that I was targeted as a Jew and frequently physically and verbally attacked by other kids.

As an adult in grad school and a professor in medical school, stress became an everyday affair for years, as I struggled to get research grants in a highly competitive field and similarly in trying to publish my research in another highly competitive endeavor. On a personal level, my marriage became another major life stress. Collectively, the anger I experienced was a root cause that led to the removal of my dysfunctional gall bladder, which revealed no pathology and was simply deemed nonfunctional.

I think my chosen career perfectly illustrates the concepts of "good" stress (eustress) and "bad" stress (distress), first defined by Hungarian endocrinologist Hans Selye in 1976. Eustress provides the challenges that make life worth living. "This action-enhancing stress gives the athlete the competitive edge and the public speaker the enthusiasm to project optimally," writes Salleh.[5] I've never been lacking enthusiasm, and I've never had stage fright. In fact, before my diagnosis, I would have said that in my career I was thriving on eustress and had been doing so ever since my first microscope in second grade revealed an invisible, vibrant world smaller than I was. When I graduated to an electron microscope, I was excited to go to work every day to view that microscopic world in more detail, realizing that every day I was seeing a world no one had ever seen before. My research life was an exciting, *Star Trek*–like experience, except that instead of going into deep space, my journeys took me deep into the normally invisible structure and lives of cells.

But my commitment to an honest lifestyle inventory forces me to concede that bringing my work to others also includes distress-enhancing stimuli like airports, hotels, and time-change-induced sleep-pattern disruptions. I regularly give presentations to large audiences of usually more than one thousand. On the one hand, I love it, and the committed, enthusiastic, cultural creatives I meet energize me (good stress). On the other hand, it is exhausting (bad stress). For example, I recently gave a presentation to a wonderful audience of two thousand in São Paulo, so wonderful that after my more than three-hour presentation, I spent another three hours signing books.

Consider these statistics compiled by WebMD from the National Institute of Mental Health, the American Heart Association, and the Mayo Clinic:[6]

- 43 percent of all adults suffer adverse health effects from stress.
- 75–90 percent of all doctor's office visits are for stress-related ailments and complaints.
- Stress can play a part in headaches, high blood pressure, heart problems, diabetes, skin conditions, asthma, arthritis, depression, and anxiety.
- The Occupational Safety and Health Administration (OSHA) declared stress a hazard of the workplace that costs American industry more than $300 billion annually.

What is everyone so stressed out about? Well, there's paying the mortgage, fear of losing your job to artificial intelligence, getting stuck in traffic with drivers who seem to have lost their minds, fear that humans will destroy themselves and the planet (more on that in the next chapter) . . .

and I could go on. Not a saber-toothed tiger in sight, but still, many of us are stuck in a chronic protection mode. So much so that the authors of a new study argue that in the United States, "most daytime hours of healthy working-age adults are spent in moderate threat arousal."[7]

One added note about fight or flight. Recent research has augmented the understanding of how the body handles stress so much that scientists believe that "fight or flight" should be considered a metaphor, not a literal description. For example, they note that there should be a third concept added to fight or flight, which is "freeze," an important concept for juries and rape victims to understand: "Just because somebody didn't fight back doesn't mean they weren't assaulted. When our nervous systems sense profound danger, we often respond by freezing, shutting down, and dissociating. The response is not voluntary but autonomic (again: meaning automatic)" write Stephen W. and Seth Porges in their book *Our Polyvagal World: How Safety and Trauma Change Us.*[8]

Chapter 16

PUSHING BACK THE DOOMSDAY CLOCK

In 1945 Albert Einstein, J. Robert Oppenheimer, and University of Chicago scientists who helped develop the first atomic weapons in the Manhattan Project founded the *Bulletin of the Atomic Scientists.* Two years later the *Bulletin* unveiled its Doomsday Clock, the now-familiar image of a countdown clock with midnight signifying global apocalypse. Over the last decades, the hands of the annually adjusted clock moved backward when the world seemed to be moving toward peace, the farthest back in 1991. With the end of the Cold War, the United States and the Soviet Union signed the Strategic Arms Reduction Treaty, prompting the *Bulletin* to set the clock hand to seventeen minutes to midnight. In other years, the hands were set closer to midnight, for example, during the Vietnam War in the late sixties and in 1974 and 2007 when India and North Korea conducted their first nuclear-device testing.

More recently, the threats factored into the position of the clock hands have expanded to include climate change, biological weapons, AI-designed military weapons, AI-empowered disinformation, and emerging and reemerging infectious diseases, a formidable list that resulted in the hands being placed at eighty-nine seconds to midnight as of early 2025, the closest they had been since 1947. (As this book goes to press in 2026, the hands have been placed even closer to 85 seconds before midnight.) The *Bulletin's* statement explaining the draconian placement ended with this strongly worded paragraph:

> Blindly continuing on the current path is a form of madness. The United States, China, and Russia have the collective power to destroy civilization. These three countries have the prime responsibility to pull the world back from the brink, and they can do so if their leaders seriously commence good-faith discussions about the global threats outlined here. Despite their profound disagreements, they should take that first step without delay. The world depends on immediate action. It is 89 seconds to midnight.[1]

I can't argue with the eighty-nine seconds, and I've never been one to shy away from talking and writing about *Homo sapiens'* self-destructive practices. There is no doubt that human disruptive behaviors are the primary source contributing to the breakdown of the web of life. But I've also studied the impact of stress, and I know that it is not only individual, modern-day fears that I talked about in the last chapter that can depress the immune system, but also the constant drumbeat of stories about environmental destruction and warnings that humans may go the way of dinosaurs are taking a toll, especially on young people.

In a 2023 editorial in *Epidemiologia*, Dr. Arnaud Chiolero, professor of population health at the University of Fribourg, cited a large, international, "poignant" study of ten thousand young people (ages sixteen through twenty-four). Sixty-seven percent reported being sad, 51 percent helpless, and 61 percent powerless about climate change. "Even more striking was that 56% considered humanity to be doomed and 39% reported being hesitant to have children." In the editorial, Chiolero argues that "researchers should be wary of how they can contribute to eco-anxiety by the way they frame their research and their narratives of the evidence."[2]

I am wary about adding to eco-anxiety. I am also an optimist and my theory of conscious evolution underscores that optimism because I believe that we live in a fractal Universe, and that the self-repeating patterns of evolution point the way to a higher, harmonious level of evolution for *Homo sapiens*, though I know that's sometimes difficult to believe, surrounded as we are by chaos. My understanding of our fractal Universe means that I believe we can evolve to a higher level of evolution. As for the planet, in the spirit of that optimism, I'd like to offer some good news. Even though human-induced environmental damage has become so extreme that many fear the planet will not survive, once humans are out of the equation, Nature makes dramatic recoveries.

For example, following the Chernobyl nuclear disaster in 1986, the high level of environmental radiation led to the exclusion of humans in a large zone. It was estimated that the exclusion zone would likely not be habitable for humans for at least twenty thousand years. However, within three years, the former city began to recover. Trees started growing everywhere in the city of Pripyat, including out of high-rise buildings' windows and every floor from the ground to the roof. The newly created forests around Chernobyl are inhabited by a higher density of wild animals—including wolves, Eurasian lynx, European bison, bears, and wild boars—than anywhere else in the former Soviet Union.[3]

Another example is New Zealand's forty-four marine reserves and high-protection areas where fishing is restricted or prohibited. The first reserve was established in Goat Island Bay in 1977. At the time, marine lab divers noticed that large swaths of reef in Goat Island Bay were barren. The seaweed bed that formerly provided a rich ecosystem for thousands of marine species was grazed to bristlelike stalks by kina, a species of sea urchins. The population of the kina grazers had exploded because their chief predators, snapper and spiny rock lobsters, had been virtually fished out.

When the reserve introduced a ban on fishing, the predator and prey imbalance began to reverse. Kina populations dropped, kelp beds grew back, and snapper once again became abundant. The fish, shrimp, and lobster populations inside the reserve grew to about fifteen times higher than outside the reserve. Word of this ecological revival soon spread, and the world has beaten a path to Goat Island's shore. Today, a hundred thousand people a year visit the bay to observe the abundant fish population.

Within forty years, all of the marine reserves have experienced a significant recovery with increases in biomass and dense fish populations. Their success has created a "spillover effect," wherein fish from inside the reserve serve to replenish stocks in the surrounding, unprotected waters. Once protected from human activity, the marine reserves have illustrated that the sea, within a very short time, can return to the rich ecosystem that existed before humans arrived. The recovery has become a model for the world's environmentalists.[4]

A further demonstration of the resilience of Nature is an Indonesian island that was covered by lava flows from an eruption of Krakatoa in 1960. All life-forms on the island before the volcanic eruption were wiped out, leaving what scientists referred to as a state of "sterilization." Within five decades following the eruption, the island had become an ecological paradise, recolonized by Nature with an incredibly diverse biosphere of trees, shrubs, and plants.[5]

Can we, like Nature, heal? I believe we can. There is no doubt that human hubris has engaged misperceptions and destructive technologies that undermine the web of life and led to the setting of the Doomsday Clock at eighty-five seconds to midnight. But the clock's hands have moved backward in the past, and they can move backward again.

Evolution has given us many gifts, including consciousness. Let's use our highly evolved consciousness to create a sustainable society in which we recognize that we and the planet are literally all in this together in an energetic Universe. We can emerge from the self-destroying and planet-destroying Darwinian nightmare and move UP to the next level of evolution, a superorganism called Humanity, a multicellular organism bound together by love for each other and for Nature, of which we are part.

EPILOGUE

WHY IS THIS RELEVANT?

When I was a medical school professor, my lectures were received by students as statements of FACT. So much so that they would studiously write down every fact I spouted. Their idea was that in order to pass the course, they would simply memorize the data provided in the lecture hall. Utilizing rote memorization, my students downloaded the facts and hoped they would be able to repeat them at exam time.

At exam time, each member of the team of professors teaching histology, cell biology, and embryology would submit a series of questions that were incorporated into a final exam. At some point, I became aware that the students were very angry with me because most could not pass the questions on my section of the exam. To alleviate the problem, the following year, I told the students that my questions would relate to the information on page twelve and page twenty-nine in the coursebook we provided to the students. Again, they failed and were still angry about the questions I offered. The following year, I told the students that the final exam question was based on the BOLD-printed sentences on pages twelve and twenty-nine. Again, they failed.

I could not understand what the problem was. Then a student came forward and said he knew why the class failed my part of the exam: "The problem is that you want the students to think." I realized that my questions were designed to *connect* the relationship between the point made on page twelve and the point made on page twenty-nine, whereas the students were expecting to memorize and then simply regurgitate the facts on those pages.

It had become apparent that rote memorization of facts did not in any way help integrate the meaning of those facts and the role they played in the functioning of the body's systems, so I changed the format of my lectures. I began to present important facts and would repeatedly follow up by asking, "Why is this relevant?" Then I would reveal how the facts could be integrated into an understanding of biology. By repeatedly answering the question "Why is this relevant?" I was emphasizing that memorizing facts was not enough. To truly understand the coursework, students needed to understand how each fact played a vital role in the behavior of the body.

My current work on evolution theory described in this book provides an understanding of the mechanisms of conscious evolution. The compiled data that I've provided reveals how consciousness is the driving force behind human evolution. It is now an appropriate moment to connect the dots by answering the question I asked my students over and over: "Why is this relevant?"

1. In a world based on Newtonian physics, the substance of the Universe was divided into matter and energy, which were understood to be separate, non-interactive realms. This notion led to a division between the intangible energy fields of consciousness and spirituality and the physical realm. Accordingly, it was thought that the physical body is influenced only by physical things (e.g., pharmaceuticals and surgery) and not by elements of the energy realm (beliefs and energy signals).
2. In 1927, quantum physics established that physical atoms are actually a unique form of "focused" energy. Consequently, the Universe is comprised of only one realm: an interactive matrix of energy vibrations. Einstein's quote, "The field is the sole governing agency of matter," emphasizes that invisible

energy fields, including consciousness and spirituality, play a primary role in the unfolding of life.

3. Darwinian theory, which emphasizes the role of gene complexity and the role of competition in the evolution of life, was found to be flawed when the human genome project revealed that humans have about the same number of genes as primitive organisms found near the base of the Tree of Life.
4. The new science of epigenetics emphasizes that the genetic activity, behavior, and fate of living organisms are directly influenced by their interaction with the environment. Human beings were not a separate creation from the web of life but instead evolved from the early planetary environment, which may be perceived as a lush "Garden," and designed to live in *harmony with* not *dominion over* nature. We are an evolutionarily advanced organism here to serve as "gardeners" tending to nature's harmony.
5. Having lost our connection with Nature, human civilization has undermined the Garden and is now facing an imminent collapse. The reason for our self-destruction is that to be sustainable, our current human civilization needs the resources of 1.6 planet Earths. Given that there is only one planet Earth, humans are pillaging the planet's resources and destroying the ecosystem to acquire the extra 0.6 planetary resources. The pillaging must be reversed; to

survive as a species, humans must rapidly and radically change the way we are living.

6. Quantum science emphasizes that life is controlled by the mind and spirituality. Each human has the power to be the creative force in shaping the character of their own lives, the role provided by the conscious mind. However, 95 percent of a person's behavior and character is controlled by their programmed subconscious mind, which is simply the equivalent of a "hard drive" with downloaded habits (i.e., programs) that shape the character of a society.
7. Over 60 percent of human subconscious programs (habits) consist of disempowering, self-sabotaging, and limiting programs that are passed down from generation to generation in a child's first seven years of life. That makes most (95 percent) of an individual's behavior driven by downloaded subconscious habits derived from the behaviors of others (primarily parents and local community). The underlying problem is that individuals are unaware that they are unconsciously creating the character of their lives with the programs in their subconscious mind.
8. To survive, civilization must become aware of an important fact: Our true source/identity is the expression of a personalized and *immortal* energy "field" aka Spirit. All humans come from the SAME source; it is only cultural programming

that is responsible for the divisiveness and disempowerment that people experience. Conflicting developmental programs downloaded into the minds of children are the primary cause of civilization's self-destructive behaviors and global chaos that is undermining human survival.

9. Awakening to the facts provided in this book and incorporating them into our collective behavior will enable humanity to evolve from lives based on competition and struggle to lives based on cooperation, harmony, and love.

As history reveals, the end of a civilization is NOT the end of humanity; civilizations come and go. As one civilization breaks down into chaos, it offers the opportunity for individuals to come together and manifest a more advanced and empowering version of civilization. This is not a suggestion but an awareness that stems from identifying fractal iterations that have come before and will continue to repeat into the future.

Futurist Buckminster Fuller emphasized, "*You never change things by fighting the existing reality. To change something, build a new model that makes the existing model obsolete.*" It is my sincerest hope that the information provided in this work can serve as a foundation for a model of new version of civilization. So, my answer to the question I repeatedly asked my students, "Why is this relevant?" is that this work can be considered a “foundation” for a more inclusive theory, one that can be modified and adjusted to seed the next step of evolution that will enable a new, holistic, and harmonious version of civilization.

With best wishes for yours and the planet’s health, harmony, and happiness,

Bruce H. Lipton

Acknowledgments

The scientific inspiration and the creation of this book have been guided by both spiritual and incarnate muses. I acquired knowledge from many teachers, muses to whom I am particularly indebted. I wish to acknowledge the generous support of the following muses who have helped make this book a reality.

I am beholden to the *spirits* of science as academic muses, for I am fully aware that *forces* outside of myself have guided me in bringing this message to the world. The specific scientific endeavors that guided the research in this book were primarily inspired by the lives, works, and spirits of three distinguished and prominent scientists: Jean-Baptiste Lamarck, Albert Eistein, and Harold Morowitz.

ACKNOWLEDGMENTS

The science research that is the foundation of this book has been encouraged by my most wonderful life and business partner, Margaret M. Horton. Margaret's love has made her a guiding muse who always supported the time and effort spent in manifesting our collective vision in support of human evolution.

It is my hope that this book will enhance the lives of all humans, with my special attention to the support offered by my family: my daughter Tanya, her son Gabriel, and daughter Lily-Annabel; my daughter Jennifer, her partner Stef McKinstray, and their son, Miles; and my brother David, his wife Cindy, and their son Alex.

In the sense that "it takes a community," I am truly indebted to the staff of Mountain of Love Productions, my extended family: President and founder Margaret Horton; Annie Willis, the best Business Manager ever and a great friend; Laura Virgallito-Revere, dear friend and Social Media manager; Beth Karish, keeping the peace as Customer Service manager; Marco and Kayla Morelli of Cosmos Cooperative, our fabulous Web Masters who keep the message of the New Science radiating out into the world; Cindy Lipton, Fulfillment Manager servicing our website store; and Robert Mueller, my spiritual son, illustrator, and manuscript reviewer. I am honored by the support offered by my nephew Alex Lipton, producer-videographer of hundreds of my presentations, as well as the protégé I hope will expand on the new evolution science.

ACKNOWLEDGMENTS

The intention of this work is to awaken the public to their role in manifesting a new, sustainable civilization. Toward that end, I offer my gratitude for the collegial support from my dear brother Gregg Braden; Sue Morter, PhD; Anita Moorjani; Shamini Jain, PhD; and Lynne and Bryan McTaggart.

My message of creating a future in which we all thrive has been delivered to worldwide audiences through the supportive efforts of wonderful conference producers who have given me multiple opportunities to speak, including: Pablo and Robin Sutterlin (Younity), Richard Smith (TCCHE), Elan Cohen (Shaloha Productions), Jorge Patrono (Creando tu vida), Liz Dawn (Celebrate Your Life), Alun Biggart (Alun.dk), Burgi Sedlak (Prana Vita), Jane Stephenson (Seminars.ie), and Dra. Renata Eleandro (Fisioquantic). In this vein, I must also thank the networks that distribute my programs: Gaia, Humanity's Team, Younity, Hay House, Sounds True, and the Shift Network.

I am honored to have editor Patricia King as a colleague in creating this book. Patricia is a Bay Area freelance writer and former *Newsweek* reporter who worked for a decade as the magazine's San Francisco Bureau Chief. Patricia's amazing skills at integrating, editing, and synthesizing information are directly responsible for the clarity of this book. She accomplished miracles in downloading and understanding the New Science to the extent that she was even able to expand on its topics.

ACKNOWLEDGMENTS

Exceptional help in manifesting this book is due to the support of BenBella Books, Glenn Yeffeth, and his most amazing staff. BenBella is the best publishing agency I have ever experienced; their support and encouragement have made the process of manifesting a book an author's dream. A special-thank you to our agent, Anne Archer, an integral part of our team who brought us to BenBella Books.

The New Science described in this book emphasizes the role of energy vibrations in shaping the material world. In the world of quantum physics, good "vibes," scientifically referred to as constructive interference, are powerful influences in shaping the character of our lives. I have been continuously encouraged and uplifted by the music. I have been especially honored by my connection with musicians Lauren Monroe and her partner, Richard Allen; Jon Anderson and his partner, Jane; Jonathan Goldman and his partner, Andi; Deva Premal and Miten; Kristin Hoffman; and Dr. Rev. Paul Luftenegger, all loving people who spread that love around the world.

Last, but not least, I want to acknowledge and thank the readers and audience of people who have followed the science of *The Biology of Belief* and *Spontaneous Evolution*. It is their support that enabled my message of self-empowerment to touch the minds and hearts of cultural creatives and seekers around the globe. Everyone in this audience is an active participant in the advancement of an imminent evolution.

Notes

PART 1

1. R. C. Henry, "The Mental Universe," *Nature* 436, 7 (2005): 29.
2. M. Gray, "Lynn Margulis and the endosymbiont hypothesis: 50 years later," *Mol Biol Cell* 28, 10 (2017): 1285–1287.
3. Lynn Margulis and Dorion Sagan, *Origins of Sex: Three Billion Years of Genetic Recombination* (New Haven, CT: Yale University Press, 1986).

CHAPTER 1

1. Gray, "Lynn Margulis," 1285–1287.
2. R. A. Slotten, *The Heretic in Darwin's Court* (Columbia University Press, 2004).
3. Harold Morowitz, *Beginnings of Cellular Life: Metabolism Recapitulates Biogenesis* (Yale University Press, 1992).
4. M. Balter, "Was Lamarck Just a Little Bit Right?," *Science* 288, 5463 (2000): 38.

5. Peter Ward, *Lamarck's Revenge: How Epigenetics Is Revolutionizing Our Understanding of Evolution's Past and Present* (Bloomsbury Publishing, 2018).
6. B. Worm, "A most unusual (super)predator," *Science* 349, 6250 (2015): 784–785.
7. World Wildlife Fund, "What Is the Sixth Mass Extinction Event and What Can We Do About It?," https://www.worldwildlife.org/resources/explainers/what-is-the-sixth-mass-extinction-and-what-can-we-do-about-it.
8. S. Motesharrei, et al., "Human and nature dynamics (HANDY): Modeling inequality and use of resources in the collapse or sustainability of societies," *Ecological Economics* 101 (2014): 90–102.

CHAPTER 3

1. Connie Barlow, *From Gaia to Selfish Genes, Selected Writings in the Life Sciences* (MIT Press, 1992), 228.
2. D. G. Booth, et al., "3D-CLEM Reveals That a Major Portion of Mitotic Chromosomes Is Not Chromatin," *Mol Cell* 64, 4 (2016): 790–802; G. M. Cooper and K. Adams, *Chromosomes and Chromatin*, 2nd Edition, (Oxford University Press, 2000).
3. J. Cairns et al., "The Origin of Mutants," *Nature* 335 (1988): 142–145.
4. T. Horinouchi and C. Furusawa, "Understanding metabolic adaptation by using bacterial laboratory evolution and transomics analysis," *Biophysical Reviews* 12, 3 (2020): 677–682.
5. L. López-Maury, et al., "Tuning gene expression to changing environments: From rapid responses to evolutionary adaptation," *Nature Reviews Genetics 9* (2008): 583–593.
6. A. Shi, et al., "Microbial adaptive evolution," *J Ind Microbiol Biotechnol* 49, 2 (2022): kuab076.
7. P. L. Foster, "Adaptive mutation: implications for evolution," *Bioessays* 22, 12 (Dec 2000): 1067–1074.
8. C. Y. Ji et al., "Detecting Potentially Adaptive Mutations from the

Parallel and Fixed Patterns in SARS-CoV-2 Evolution," *Viruses* 14, no. 5 (May 18, 2022): 1087.

9. B. H. Lipton, "A fine structural analysis of normal and modulated cells in myogenic culture," *Developmental Biology* 60 (1977): 26–47.
10. B. H. Lipton, et el., "Histamine-Modulated Transdifferentiation of Dermal Microvascular Endothelial Cells," *Experimental Cell Research* 199 (1992): 279–291.
11. H. F. Nijhout, "Metaphors and the Role of Genes in Development," *Bioessays* 12, 9 (1990): 441–446.

CHAPTER 4

1. J. N. Timmis, et al., "Endosymbiotic Gene Transfer: Organelle Genomes Forge Eukaryotic Chromosomes," *Nature Reviews Genetics* 5 (2004): 123–135.
2. N. C. Rockwell, et al., "Primary endosymbiosis and the evolution of light and oxygen sensing in photosynthetic eukaryotes," *Front Ecol Evol* 2, 66 (2014); T. Cavalier-Smith, "Origin of mitochondria by intracellular enslavement of a photosynthetic purple bacterium," *Proc Biol Sci* 273, 1596 (2006): 1943–1952.

CHAPTER 6

1. C. R. Woese, *The Genetic Code: The Molecular Basis for Genetic Expression* (New York: Harper & Row, 1967); M. Yarus et al., "RNA–Amino Acid Binding: A Stereochemical Era for the Genetic Code," *Journal of Molecular Evolution* 69 (2009): 406–429.
2. Yarus, et al., "RNA–Amino Acid Binding," 406–429.
3. Woese, *The Genetic Code*; Yarus, et al., "RNA–Amino Acid Binding," 406–429.

CHAPTER 7

1. Morowitz, *Beginnings of Cellular Life.*

2. The Physics arXiv Blog, "First evidence of cell membrane molecules in space," *Astronomy Magazine*, May 28, 2021.
3. Morowitz, *Beginnings of Cellular Life.*

CHAPTER 8

1. Albert Einstein, Condolence letter to the family of Michael Besso, March 21, 1955.

CHAPTER 9

1. S. Miller and H. Urey, "A Production of Amino Acids Under Possible Primitive Earth Condition," *Science* 117 (1953): 528–5293.
2. D. P. Glavin, et al., "Abundant ammonia and nitrogen-rich soluble organic matter in samples from asteroid (101955) Bennu," *Nature Astronomy* 9 (2025): 192–210.
3. C. Chyba and C. Sagan, "Endogenous production, exogenous delivery, and impact-shock synthesis of organic molecules: an inventory for the origins of life," *Nature* 355, 6356 (1992): 125–132.
4. Y. Kawaguchi, et al., "DNA Damage and Survival Time Course of Deinococcal Cell Pellets During 3 Years of Exposure to Outer Space," *Frontiers in Microbiology* 11 (2020): 2050.
5. Nick Lane, *Transformer: The Deep Chemistry of Life and Death* (W. W. Norton & Company, 2022), 113.
6. Lane, *Transformer*, 158–159.
7. W. Martin et al., "Hydrothermal Vents and the Origin of Life," *Nature Reviews Microbiology* 6 (2008): 805–814.
8. M. Van Kranendonk, et al., "Life Springs," *Sci Am* 317 (2017): 28–35.
9. T. Djokic, et al., "Earliest signs of life on land preserved in ca. 3.5 Ga hot spring deposits," *Nat Commun* 8, article 15263 (2017).

CHAPTER 10

1. B. Sapoval, et al., "Fractal interfaces in the self-stabilized etching of random systems," *Europhys. Lett.* 41, 3 (1998): 297–302.

NOTES

CHAPTER 11

1. R. J. Baumgartner, et al., "Nano-porous pyrite and organic matter in 3.5-billion-year-old stromatolites record primordial life," *Geology* 47, 11 (2019): 1039–1043.
2. M. Starr, "Gobsmacking Study Finds Life on Earth Emerged 4.2 Billion Years Ago," *ScienceAlert*, July 12, 2024, https://www.sciencealert.com/gobsmacking-study-fines-life-on-earth-emerged-4-2-billion-years-ago.

CHAPTER 12

1. W. C. Gough, "The Cellular Communication Process and Alternative Modes of Healing," *Subtle Energies & Energy Medicine* 8, 2 (1997): 92–93.

PART 2: INTRODUCTION

1. Bruce H. Lipton, *The Biology of Belief, 10th Anniversary Edition: Unleashing the Power of Consciousness, Matter and Miracles* (Carlsbad, CA: Hay House, 2016); Marianne Szegedy-Maszak, "Mysteries of the Mind: Your Unconscious Is Making Your Everyday Decisions," *U.S. News & World Report* (2005).
2. Sarah Knapton, "Nine in 10 Cancers Caused by Lifestyle," *The Telegraph*, December 17, 2015.
3. Dominik Nischwitz, *It's All in Your Mouth: Biological Dentistry and the Surprising Impact of Oral Health on Wellness* (Chelsea Green Publishing, 2020), 168.

CHAPTER 13

1. Ariana Eunjung Cha, et al., "Iowa's great cancer mystery: Is the land making young people sick?," *The Washington Post*, October 28, 2025.
2. "Exploring Monocultures and 'Green Deserts,'" *Planet Wild*, May

22, 2023, https://planetwild.com/blog/monoculture-green-deserts-and-5-ways-you-can-help#:~:text=One%20of%20the%20biggest%20issues%20with%20monoculture,makes%20it%20harder%20for%20plants%20to%20grow.
3. G. Bryja, "Predator Killing: Greed, Savagery, and Misguided Science," *Omere,* Dec 26, 2023.
4. D. Bloch, "The whale pump: Investigating humpback whale-driven nutrient cycling in Southeast Alaska," *Alaska Whale Foundation*, June 8, 2023.
5. World Health Organization, "Obesity and Overweight," December 8, 2025, https://www.who.int/news-room/fact-sheets/detail/obesity-and-overweight.
6. J. Sanford, "Ultra-processed food: Five things to know," *Stanford Medicine*, July 25, 2025.
7. American Dental Association to Food and Drug Administration, letter, October 20, 2025, re: Docket No. FDA-2025-N-1793, "Ultra-Processed Foods; Request for Information."
8. Sanford, "Ultra-processed food: Five things to know."
9. Chris Van Tulleken, *Ultra-Processed People: Why We Can't Stop Eating Food That Isn't Food* (W. W. Norton & Company Ltd., 2023).
10. Van Tulleken, *Ultra-Processed People.* https://www.amazon.com/Ultra-Processed-People-Cant-Stop-Eating/dp/1324076267.
11. Sanford, "Ultra-processed food: Five things to know."
12. K. D. Hall, et al., "Ultra-processed diets cause excess calorie intake and weight gain: An inpatient randomized controlled trial of ad libitum food intake," *Cell Metab.* 30, 1 (2019): 67–77.
13. Break Free From Plastic, Brand Audit Report, 2023.
14. Michael Pollan, *In Defense of Food: An Eater's Manifesto* (Penguin Press, 2008).
15. Sanford, "Ultra-Processed Food: Five things to Know."

16. L. K. Loch, et al., "Ultra-processed food addiction in a nationally representative sample of older adults in the USA," *Addiction*, 2025.
17. Van Tulleken, *Ultra-Processed People.*
18. E. Crosbie, et al., "A policy study on front-of-the-pack nutrition labeling in the Americas: emerging developments and outcomes," *The Lancet Regional Health—Americas*, 18 (2023): 100400.
19. E. Crosbie, et al., "Implementing front-of-pack nutrition warning labels in Mexico: important lessons for low- and middle-income countries," *Public Health Nutrition* 26, 10 (2023): 2149–2161.
20. L. Wang, et al., "We Are What We Eat: The Association Between Ultra-Processed Food Consumption, Colon Cancer Risk and All-Cause Mortality," *BMJ* 2022, 378 (2022): e068921.
21. On Yee Annie Chan et al., "The Association of Dietary Fiber Intake with Colorectal Cancer and Related Risks: A Literature Review of Recent Research," *Journal of Agriculture and Food Research*, 21 (2025): 10199.
22. B. Chassaing, et al., "Randomized Controlled-Feeding Study of Dietary Emulsifier Carboxymethylcellulose Reveals Detrimental Impacts on the Gut Microbiota and Metabolome," *Gastroenterology* 162, 3 (2022 March): 743–756.
23. M. M. Lane, et al., "Ultra-processed food exposure and adverse health outcomes: umbrella review of epidemiological meta-analyses," *BMJ* 384 (2024): e077310.
24. T. M. Hsu et al., "Effects of sucrose and high fructose corn syrup consumption on spatial memory function and hippocampal neuroinflammation in adolescent rats," *Hippocampus* 25, 2 (Feb 2015): 227–39.
25. W. Song, et al., "Dietary Fibre-Adapted Gut Microbiome Clears Dietary Fructose and Reverses Hepatic Steatosis," *Nature Metabolism* 7 (Sept 2025): 1801–1818.
26. Maria Balhara, "'Gateway foods' like candy, pastries and frozen treats may lead to unhealthy teen eating." Presented at the American Heart Association's Hypertension Scientific Sessons, San Diego, September 7–10, 2022.

CHAPTER 14

1. Dominik Nischwitz, *It's All in Your Mouth: Biological Dentistry and the Surprising Impact of Oral Health on Wellness* (Chelsea Green Publishing, 2020), 168.
2. Nischwitz, *It's All in Your Mouth*, 168.
3. Robert Kulacz and Thomas E. Levy, *The Roots of Disease: Connecting Dentistry & Medicine* (2002).
4. D. Fu, et al., "Connection between oral health and chronic diseases," *MedComm* 2025, 6: e70052.
5. S. Sen, et al., "Gum Disease and Cavities Linked to Increased Stroke Risk," *Neurology Open Access*, October 22, 2025.
6. Nischwitz, *It's All in Your Mouth*, 168.
7. J. J. Rajasekaran et al., "Oral Microbiome: A Review of Its Impact on Oral and System Health," *Microorganisms* 12, no. 9 (August 29, 2024): 1797.
8. Nischwitz, *It's All in Your Mouth*, 168.
9. Martin J. Blaser, *Missing Microbes: How the Overuse of Antibiotics Is Fueling Our Modern Plagues* (Henry Holt, 2014).
10. Rajasekaran, et al., "Oral Microbiome."
11. X. Yuan, et al., "Systemic antibiotics increase microbiota pathogenicity and oral bone loss," *International Journal of Oral Science* 15 (2023): 4.
12. A. Amato, "Oral Microbiota, Bacterial Infections, Antibiotic Prescriptions, and Antimicrobial Resistance in Children," *Microorganisms* 11, 1927 (2023).
13. Nischwitz, *It's All in Your Mouth*, 168.

CHAPTER 15

1. R. Nelson, "Fight or Flight Theory and the Autonomic Nervous System," *APA Magazine*, 58, 3 (2025).
2. Mohd Razali Salleh, "Life Event, Stress and Illness," *Malaysian Journal of Medical Sciences* 15, no. 4 (October 2008): 9–18.
3. Salleh, "Life Event, Stress and Illness," 9–18.

4. Elena Hinz et al., "Childhood Allostatic Load Predicts Cardiometabolic Health in Adulthood," *Proceedings of the National Academy of Sciences of the United States of America* (published September 22, 2025).
5. Salleh, "Life Event, Stress and Illness," 9–18.
6. "Effects of Stress on Your Body," WebMD, February 29, 2024, medically reviewed by Melinda Ratini, MS, DO, https://www.webmd.com/balance/stress-management/effects-of-stress-on-your-body.
7. Alexandra D. Crosswell et al., "Deep Rest: An Integrative Model of How Contemplative Practices Combat Stress and Enhance the Body's Restorative Capacity," *Psychological Review* 131, no. 1 (2024): 247–270.
8. Stephen W. Porges and Seth Porges, *Our Polyvagal World: How Safety and Trauma Change Us* (W. W. Norton & Company, Inc., 2023).

CHAPTER 16

1. John Mecklin, "Closer than ever: It is now 89 seconds to midnight," 2025 Doomsday Clock Statement, January 28, 2025.
2. A. Chiolero, "Academic Framing as a Cause of Eco-Anxiety," *Epidemiologia* 2023, 4 (2023): 60–62.
3. UN Environmental Programme, "How Chernobyl has become an unexpected haven for wildlife," UNEP—UN Environmental Programme, September 16, 2020.
4. Kennedy Warne, "Blue Haven: New Zealand Marine Reserves are a model for the world," *National Geographic,* April 2007.
5. R. J. Whittaker, et al., "Plant Recolonization and Vegetation Succession on the Krakatau Islands, Indonesia," *Ecological Monographs* 59, 2 (1989): 59–123.

Index

INDEX

INDEX

INDEX

INDEX

INDEX

INDEX

INDEX

INDEX

About the Author

Photo by Gino De Grandis

Bruce H. Lipton, PhD, is an internationally recognized best-selling author and cellular biologist who has spent decades challenging scientific orthodoxy with a revolutionary message: Genes don't control our destiny—our beliefs and environment do. As a senior research scientist at Stanford University's School of Medicine, his groundbreaking cloning experiments revealed the biochemical pathways by which the mind controls behavior and genetic activity, presaging today's revolutionary field of epigenetics. Dr. Lipton's research has been published in prestigious scientific journals including *Science*, *Journal of Cell Biology*, *Developmental Biology*, *Biochemistry*, and *Experimental Cell Research.*

Translated into over 50 languages and with more than 500,000 copies sold, *The Biology of Belief* has become a global phenomenon. Dr. Lipton was named one of the world's most spiritually influential living people by the United Kingdom's *Watkins Journal* and received Japan's prestigious 2009 Goi Peace Award in honor of his scientific contribution to world harmony.